edexcel ▪▪▪

REVISE EDEXCEL GCSE

Geography A
Geographical Foundations
For the linear specification first teaching 2012

REVISION WORKBOOK

Series Consultant: Harry Smith   Author: Anne-Marie Grant

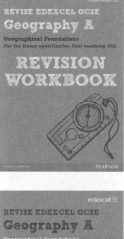

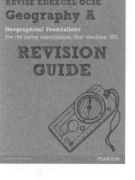

THE REVISE EDEXCEL SERIES
Available in print or online

Online editions for all titles in the Revise Edexcel series are available Autumn 2013.

Presented on our ActiveLearn platform, you can view the full book and customise it by adding notes, comments and weblinks.

Print editions

Geography A Revision Workbook   9781446905357

Geography A Revision Guide   9781446905340

Online editions

Geography A Revision Workbook   9781446905470

Geography A Revision Guide   9781446905494

Print and online editions are also available for Geography B.

This Revision Workbook is designed to complement your classroom and home learning, and to help prepare you for the exam. It does not include all the content and skills needed for the complete course. It is designed to work in combination with Edexcel's main GCSE Geography 2009 Series.

To find out more visit:
www.pearsonschools.co.uk/edexcelgcsegeographyrevision

ALWAYS LEARNING   **PEARSON**

Contents

> ✓ Make sure you know which topics you have studied – you only need to revise these.

A small bit of small print
Edexcel publishes Sample Assessment Material and the Specification on its website. This is the official content and this book should be used in conjunction with it. The questions in *Now try this* have been written to help you practise every topic in the book. Remember: the real exam questions may not look like this.

◎ **Target grade ranges**
Target grade ranges are quoted in this book for some of the questions. Students targeting this grade range should be aiming to get most of the marks available. Students targeting a higher grade should be aiming to get all the marks available.

Basic geographical skills

FOUNDN
D

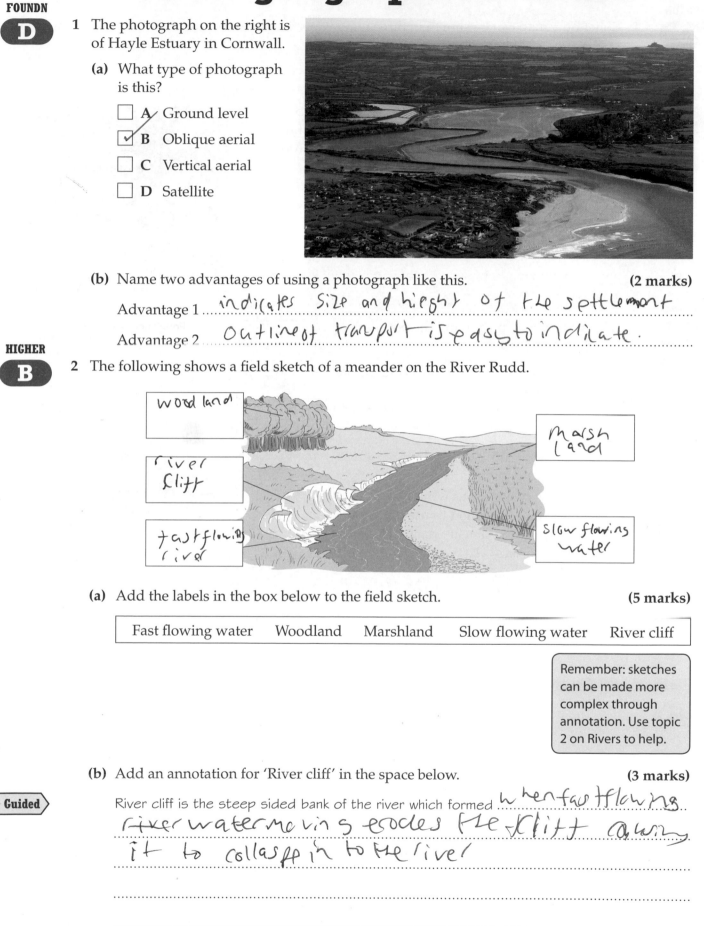

1 The photograph on the right is of Hayle Estuary in Cornwall.

(a) What type of photograph is this?

☐ **A** Ground level

☑ **B** Oblique aerial

☐ **C** Vertical aerial

☐ **D** Satellite

(b) Name two advantages of using a photograph like this. **(2 marks)**

Advantage 1 ....indicates size and hieght of the settlement....

Advantage 2 ....Outline of transport is easy to indiate.....

HIGHER
B

2 The following shows a field sketch of a meander on the River Rudd.

woodland

river cliff

fast flowing river

marsh land

slow flowing water

(a) Add the labels in the box below to the field sketch. **(5 marks)**

| Fast flowing water   Woodland   Marshland   Slow flowing water   River cliff |

> Remember: sketches can be made more complex through annotation. Use topic 2 on Rivers to help.

(b) Add an annotation for 'River cliff' in the space below. **(3 marks)**

Guided

River cliff is the steep sided bank of the river which formed ....when fast flowing....

....river water moving erodes the clitt away....

....it to collaspp in to the river....

....

....

....

1

Cartographic skills

FOUNDN
C

1 Relief on a map can be shown in three main ways. Name **two** of them.   **(2 marks)**

shading

spots hieght

FOUNDN
C

2 Find the settlement of Kingsbridge on the OS map below (starting in GR 7343). Describe the shape of the settlement.   **(2 marks)**

The settlement is nucleted.

HIGHER
A

3 (a) Study the map showing the population density for the US.

Population per square mile
250 or more ■
50–249.9 ■
10–49.9 ■
less than 10 ☐

(i) What type of mapping technique has been used to represent the information?   **(1 mark)**

choropleth mapping

(ii) Give **two** reasons why this technique is good to use for population density data.   **(2 marks)**

1   you can see the patten

2   it can be easib reconigsed.

(b) Describe the population density distribution. Use evidence from the map.   **(4 marks)**

Guided

The population density is unevenly distributed.

2

Sketch maps

> Sketch maps are simple drawings.

FOUNDN
D

1 Name **four** things that need to be added to all sketch maps.   **(4 marks)**

1 ....Key...

2 ....Sca...

3 ....tit.l...

4 ....Compass...

FOUNDN
C

Guided

2 Below is an incomplete sketch map of an island.

Read through the descriptions of the island and complete the sketch map.   **(5 marks)**

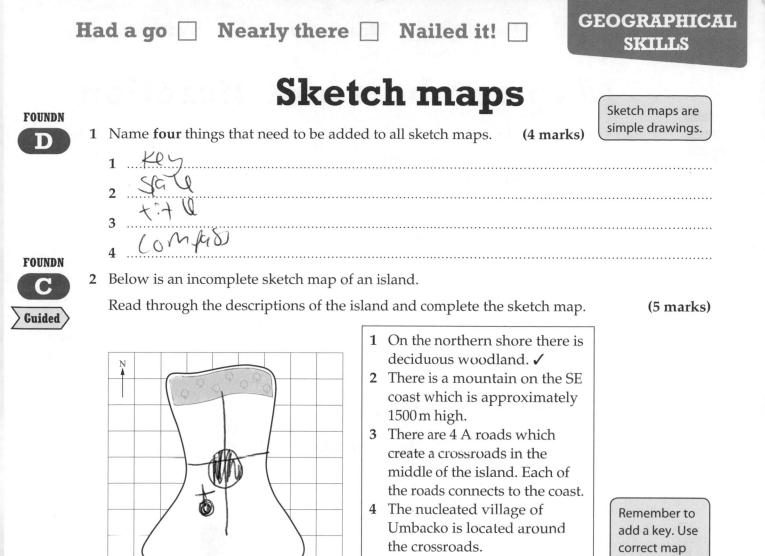

1 On the northern shore there is deciduous woodland. ✓
2 There is a mountain on the SE coast which is approximately 1500 m high.
3 There are 4 A roads which create a crossroads in the middle of the island. Each of the roads connects to the coast.
4 The nucleated village of Umbacko is located around the crossroads.
5 There is a church located just outside the SW part of the village.

> Remember to add a key. Use correct map symbols and label where appropriate.

HIGHER
C

3 Create a sketch map using the OS map extract. Include the following details:

(i)   the railway line

(ii) all roads

(iii) the settlement of Sherburn

(iv) location of church in GR 3142

(v) Broomside House   **(5 marks)**

Map symbols and direction

FOUNDN
HIGHER
C

1 (a) What do the following symbols mean?   (2 marks)

P   paiFing

✕   DICNic are

☀   view point

⬤   bus stop

> If you are not sure, look at a key on an OS map (1: 50 000)

(b) Draw the OS symbols for the following: (3 marks)

Guided

Golf course   🚩   Church with a tower   卐   Spot height   ·546

Nature Reserve   👤   Clubhouse   CH   Coniferous wood   🌲

HIGHER
C

2 Study the map extract below.

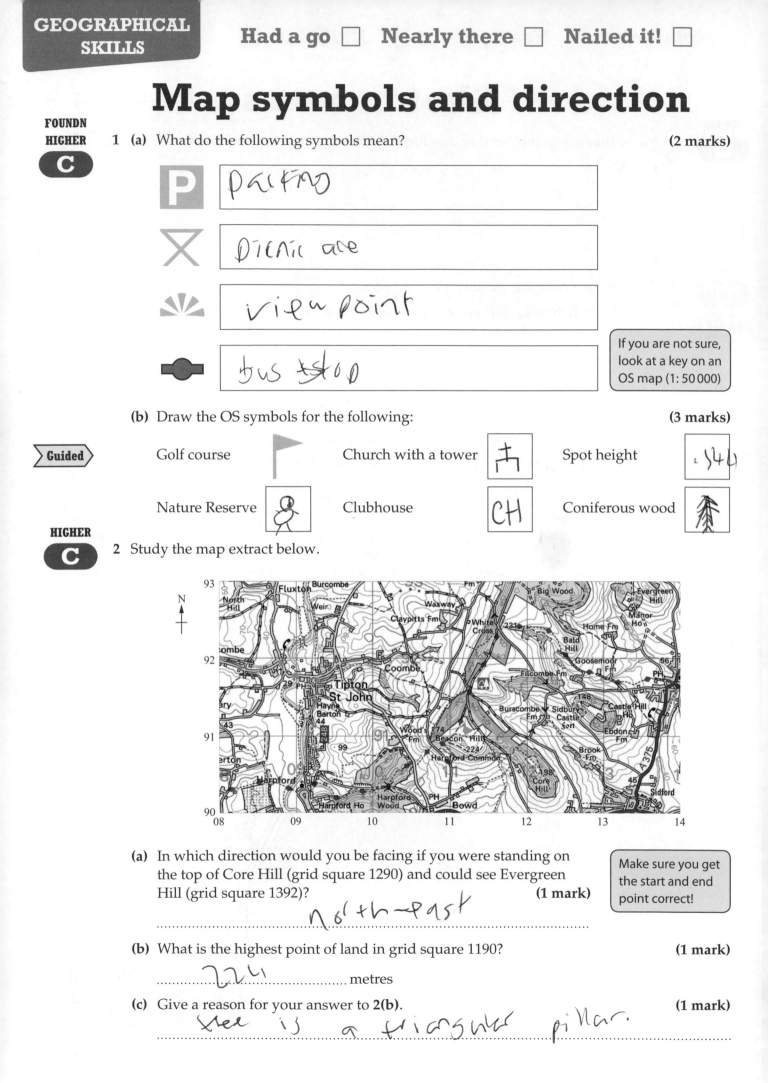

(a) In which direction would you be facing if you were standing on the top of Core Hill (grid square 1290) and could see Evergreen Hill (grid square 1392)?   (1 mark)

north~east

> Make sure you get the start and end point correct!

(b) What is the highest point of land in grid square 1190?   (1 mark)

.... 224 .... metres

(c) Give a reason for your answer to 2(b).   (1 mark)

there is a triangular pillar.

Grid references and distances

FOUNDN
D
Guided

1 Use the map extract from page 4 to help you answer the following questions.

(a) What type of woodland is shown in grid square 1292? **(1 mark)**

D_e_li_dous_ s

(b) Give the 6-figure grid reference for the nature reserve in the centre of the extract. **(1 mark)**

113911

(c) To the nearest ½ km, work out the straight line distance from the public house in grid square 1391 to the public house in grid square 0991. **(1 mark)**

4 km

> Read the question carefully!

(d) What is the name of the farm in grid square 1092? **(1 mark)**

☐ A Home Farm
☑ B Claypitts Farm
☐ C Brook Farm
☐ D Goosemoor Farm

HIGHER
C

2 There are two bridges located on the River Otter (from grid square 0990 to 0992).

What is the 6-figure grid reference of the most northerly bridge? **(1 mark)**

☐ A 099921
☐ B 095926
☑ C 091923
☐ D 095925

HIGHER
C

3 (a) To the nearest ½ km, work out the total winding distance of the River Otter. **(1 mark)**

6 km

> Remember a piece of string will help with winding distances!

(b) What symbol can be found at grid square 137912? **(1 mark)**

telephone

(c) Give the 6-figure grid reference for the highest point of land in grid square 1190. **(1 mark)**

087910 117000

(d) What is the 6-figure grid reference of the telephone which appears to be in grid square 0892? **(1 mark)**

087919

> Make sure you look carefully. This symbol has a label line!

Had a go ☐   Nearly there ☐   Nailed it! ☐

Cross sections and relief

FOUNDN
D

⟩ Guided ⟩

1 Draw lines to match up the following contour patterns (aerial view) to the cross sectional shape of the land. **(3 marks)**

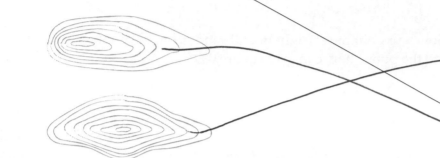
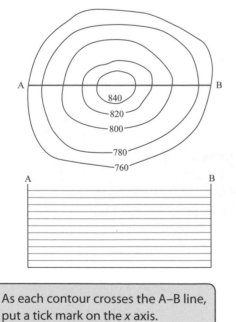

FOUNDN
D

2 Use the OS map from page 4 to help you answer the following question.

Complete the sentences by using the words in the box below. **(4 marks)**

flat	steep	99 m	199 m	flatter	steeper	40 m	70 m

The land to the east of the River Otter is .... It rises to a maximum height

of .... above sea level. In comparison the land to the west of the river is

.... and rises to a height of approximately .... above sea

level.

HIGHER
A

3 (a) Create a cross sectional diagram of the contour pattern below. **(4 marks)**

(b) Describe the relief shown by your cross section drawing. **(3 marks)**

A ———————— B

840
820
800
780
760

A ———————— B

....

....

....

....

....

....

....

....

....

> As each contour crosses the A–B line, put a tick mark on the *x* axis.

> Use a **ruler** and **sharp** pencil for all graphical skills.

Land use and settlement shapes

FOUNDN
E

1 A dispersed settlement can be described as: **(1 mark)**

☐ **A** Clustered or grouped together

☐ **B** Spread out

☐ **C** In a line along a river or road.

FOUNDN
HIGHER
C

2 Sketch the shape of:

(a) a nucleated settlement   **(1 mark)**   **(b)** a linear settlement.   **(1 mark)**

   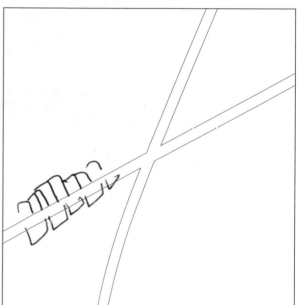

HIGHER
B

3 Use the map extract from page 4 to help you answer the following questions.

(a) Describe the shape of Sidford in grid square 1390.   **(2 marks)**

The settlement is linear

....

....

(b) Describe the physical and human land uses from grid square 1290 to grid square 1390.

Use map evidence in your answer.   **(4 marks)**

Guided

The land use is very rural. Brook Farm is located to the ....

....

....

....

....

....

....

You must include **specific** references to the map for full marks!

....

Physical and human patterns

Use the following map extract of Alnwick to help you answer the following questions.

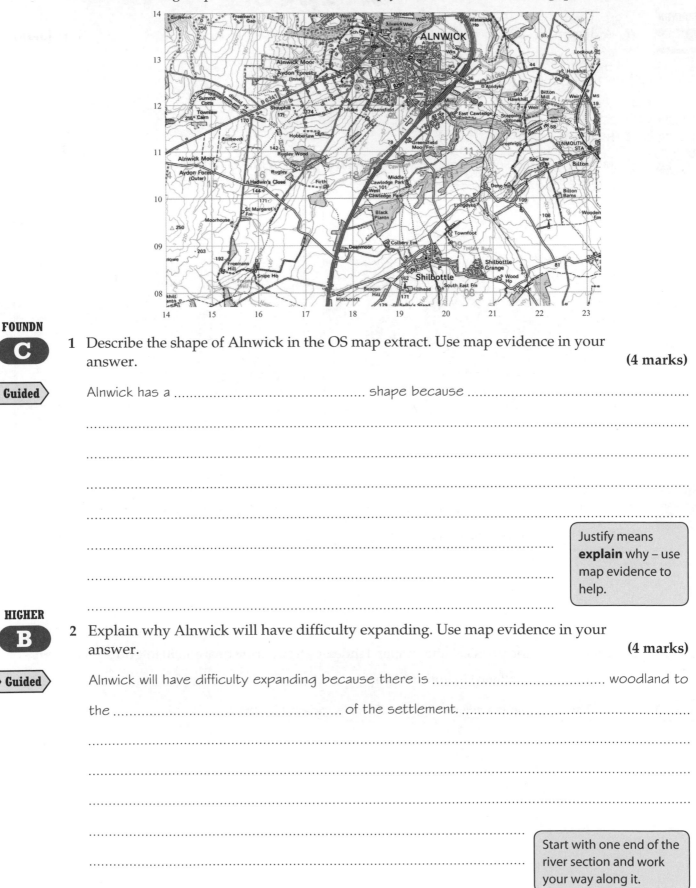

FOUNDN

C

1 Describe the shape of Alnwick in the OS map extract. Use map evidence in your answer.

(4 marks)

Guided

Alnwick has a .... shape because ....

....

....

....

....

....

....

....

> Justify means **explain** why – use map evidence to help.

HIGHER

B

2 Explain why Alnwick will have difficulty expanding. Use map evidence in your answer.

(4 marks)

Guided

Alnwick will have difficulty expanding because there is .... woodland to

the .... of the settlement. ....

....

....

....

....

....

> Start with one end of the river section and work your way along it.

Human activity on OS maps

Use the map extract of Alnwick from page 8 to help you answer the following questions.

FOUNDN
C

EXAM ALERT

1 In grid square 1912, state **two** pieces of map evidence that suggests that Alnwick attracts tourists.
(2 marks)

....

....

....

....

> Exam questions similar to this have proved tricky – be prepared! **ResultsPlus**

FOUNDN
D

2 Name **one** other piece of map evidence that proves Alnwick is popular for tourists.
(1 mark)

....

FOUNDN
C

3 Suggest what form of public transport can be used in grid square 1813.
(2 marks)

....

....

....

....

> Remember that public transport would include trains, buses and trams.

HIGHER
B

4 Shilbottle is a village to the SE of Alnwick. Give **two** pieces of map evidence which prove it's a village.
(2 marks)

....

....

....

....

> Villages will have **three** common 'services'. Some of these are closing down!

HIGHER
B

5 Shilbottle is a rural settlement. Using map evidence, justify this statement.   **(4 marks)**

Shilbottle is surrounded by a rural landscape. The many farms around the area help prove this.

Guided

For example, South East Farm located to the south east of Shilbottle. ....

....

....

....

....

....

....

Graphical skills

FOUNDN

D

1   The graph below shows the results of a 3-minute traffic survey completed by a GCSE class.

(a) Complete the graph using the data in the table.   **(2 marks)**

> **Guided**

| Lorry | 6 |
| Van | 4 |

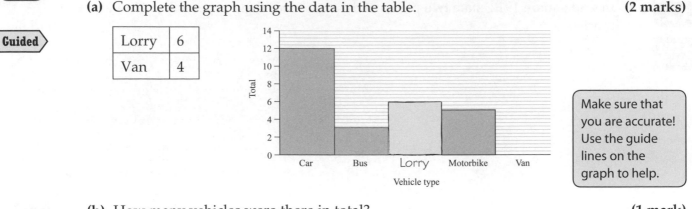

Make sure that you are accurate! Use the guide lines on the graph to help.

(b) How many vehicles were there in total?   **(1 mark)**

....

FOUNDN

HIGHER

C

2   **(a)** The graph below shows the population structure of a country. What is this type of graph called?   **(1 mark)**

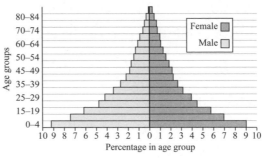

....

(b) Describe the population structure. Use evidence from the graph to support your answer.   **(4 marks)**

....

....

....

....

....

....

....

HIGHER

B

3   One pupil decides to present the traffic data in question 1 as a line graph. Explain why this graphical technique is inappropriate for the data collected.   **(3 marks)**

....

....

....

....

....

Geographical investigation

1 Give **two** advantages of using GIS to present data. **(2 marks)**

Advantage 1 ....

Advantage 2 ....

> Make sure you learn at **least** two pros and cons – these types of questions are common in all units.

FOUNDN
C

2 You have been given a task question: 'What are the impacts of tourism on Alnwick and its population?'

Describe the methods or techniques you would use to collect data to help you answer the question. **(6 marks)**

Impacts on Alnwick ....

....

....

....

....

....

Impacts on its population ....

....

....

> Think about both the positive **and** negative impacts. This links nicely to methodology in controlled assessment!

....

....

....

FOUNDN
HIGHER

3 You are using GIS to help you find out the most appropriate sites for a new wind farm.

Outline **two** pieces of information (data) you would need to add to your GIS to help you make your decision. **(4 marks)**

Wind farms are often considered ugly by locals, so it is important that the location of nearby settlements is plotted so that you can see how far they are from each proposed site. ....

....

....

....

....

....

....

....

Causes of climate change

FOUNDN
D

1 The graph shows the increase of carbon dioxide (CO_2) in the Earth's atmosphere.

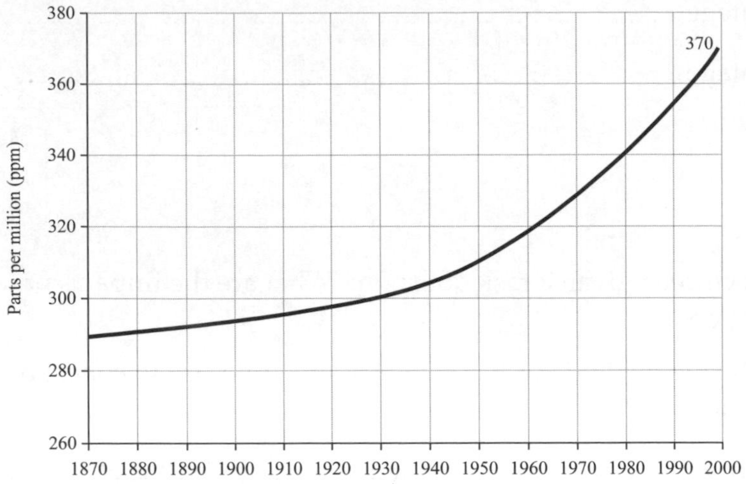

[Graph: Parts per million (ppm) on y-axis from 260 to 380; years 1870 to 2000 on x-axis. Curve rises from ~290 ppm to 370 ppm. Value 370 marked at 2000.]

How much carbon dioxide has been emitted between 1930 and 2000?   **(1 mark)**

☐ **A** 50 ppm   ☐ **B** 60 ppm   ☑ **C** 70 ppm   ☐ **D** 80 ppm

FOUNDN
HIGHER
C/B
〉 Guided 〉

2 Describe the trend shown by the graph.   **(3 marks)**

In 1750, CO_2 emissions were at their lowest with approximately 270 ppm.

....

....

....

....

....

....

> When describing patterns over **time** start at the earliest date and work your way **across**.

HIGHER
B
〉 Guided 〉

EXAM ALERT

3 Suggest **two** reasons for the possible increase in atmospheric CO_2.   **(4 marks)**

HICs are responsible for most of the CO_2 emissions due to the burning of fossil fuels (coal, oil and natural gas) to create energy for their industries and to act as fuel for

....

....

....

....

....

....

> Exam questions similar to this have proved tricky – be prepared! ResultsPlus

The negative effects of climate change

FOUNDN
D

1   The cartoon shows potential impacts from climate change. What **two** impacts are being portrayed in the cartoon? **(2 marks)**

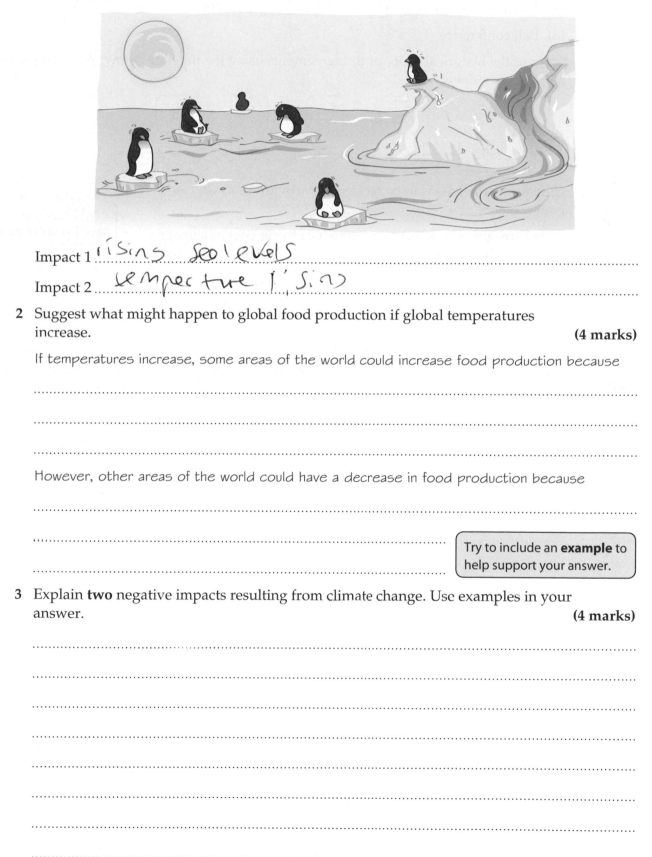

Impact 1 .....rising sea levels..

Impact 2 .....tempecture rising...

FOUNDN
C

2   Suggest what might happen to global food production if global temperatures increase. **(4 marks)**

Guided

If temperatures increase, some areas of the world could increase food production because

....

....

....

However, other areas of the world could have a decrease in food production because

....

....

....   ┌─────────────────────────────┐
                                                                                │ Try to include an **example** to │
                                                                                │ help support your answer.   │
HIGHER   └─────────────────────────────┘
B

3   Explain **two** negative impacts resulting from climate change. Use examples in your answer. **(4 marks)**

....

....

....

....

....

....

....

Responses to climate change

FOUNDN
C

1 Below are four examples of global agreements on climate change.

(a) Climate Conference, Copenhagen

(b) Kyoto, Japan

(c) Earth Summit, Rio

(d) Bali conference

Sort out the historical order of the agreements using the timeline below. An example has already been done. **(3 marks)**

☐   ☐   ☐   a

1992 ⟶ Present day

FOUNDN
HIGHER
C/B

2 Climate change needs to be tackled by individuals as well as governments.

> Think about all the ways in which energy can be saved in your house.

State **three** ways in which we can reduce the impact of climate change in the home. **(3 marks)**

1 ....

2 ....

3 ....

HIGHER
A*

Guided

3 Using examples, explain how the aims of climate conferences can be achieved on a more local scale. **(9 marks + 4 marks SPaG)**

The aims of climate conferences are primarily to reduce greenhouse gases. The most

important one being .... At a local level this

can be achieved by ....

....

....

....

....

....

....

....

....

....

....

....

Sustainable development

FOUNDN
D

1 This is the logo for the Congestion Charge. It is a way of helping transport become more sustainable.

(a) What is a congestion charge? **(2 marks)**

> **Guided**

The congestion charge is where people ....

....

....

....

(b) Name a city that uses congestion charging. **(1 mark)**

london

....

....

> There are lots of other methods that could be used. Do you know any more? Can you name an example?

HIGHER
B

2 Explain another way in which we can make transport more sustainable. Use examples in your answer. **(4 marks)**

> **Guided**

We could introduce **park and ride**, this has been successfully achieved in **Cambridge**. This will

make transport more sustainable because ....

....

....

....

....

....

....

....

Sustainable development: tropical rainforest

FOUNDN
C

1 Using an area of tropical rainforest you have studied, describe how it is being managed more sustainably. **(6 marks + 4 marks SPaG)**

Name of area ....

> **Guided**

Tropical rainforests are being sustainably managed all over the world. In ....,

they have ....

....

This is a sustainable form of management because ....

....

....

....

....

....

....

....

....

> Include relevant examples and detail to support your points. Check that your spelling, punctuation and grammar are really good, and that your answer is clear.

HIGHER
C

2 Define **sustainability**. **(2 marks)**

....

....

....

....

HIGHER
B

3 Describe **two** ways in which tropical rainforests can be managed sustainably. **(4 marks)**

....

....

....

....

....

....

....

Types of waves

FOUNDN
D

1 Waves can either be destructive or constructive. Match the statements in the box below to the correct wave type by placing the statement number in the correct box. **(2 marks)**

Destructive

3

2

Constructive

1

4

1	Low wave energy
2	Occur in stormy conditions
3	Responsible for erosion
4	Help transport material

FOUNDN
D

2 Explain what the term 'swash' means. **(1 mark)**

....

....

FOUNDN
E

3 Complete the sentences below. Use the words in the text box to help. **(5 marks)**

wind long fetch backwash powerful lots erosion deposition less short

Waves hitting the SW coast of England will have ....*powerful lots*.... of energy because

the wave has travelled a ....*long*.... distance. This means that the wave will be

very ....*powerful*.... and will cause lots of ....*erosion*.... at the coast line

because the ....*backwash*.... is greater than the swash.

> Cross through the answers in the text box once used!

HIGHER
B
Guided

4 Explain why some waves are more powerful than others. **(4 marks)**

The energy of a wave depends on three main factors – these are the fetch, speed of the
wind and the length of time the wind has been blowing. ....

....

....

....

....

....

....

....

> Attempt to explain all the factors. Include examples!

Coastal processes and landforms

FOUNDN

C

〉**Guided**〉

1 What is the difference between weathering and erosion?   **(2 marks)**

Both weathering and erosion wear away rock. The difference between them is ....

....

....

....

FOUNDN

D

2 There are four main types of erosion at the coastline. Using label lines, match the type of erosion to its definition.
                                                                                (4 marks)

Abrasion

Corrosion

Hydraulic action

Attrition

Dissolving of rocks and minerals in the cliff by sea water
Particles become rounder as they collide within water
Water which is forced into cracks and crevices in the cliff
Waves picking up rocks and hurling them at the cliff

If you are not sure about a particular one, leave it till the end!

HIGHER

A

3 Complete the diagram below by adding appropriate annotations to explain the formation of a wave-cut platform.
                                                                                (5 marks)

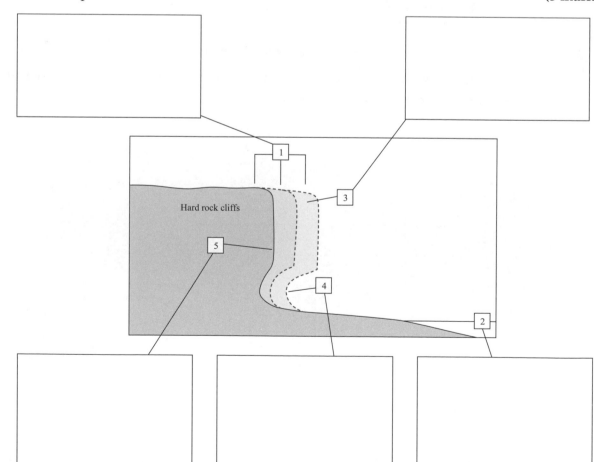

Hard rock cliffs

Erosional coastal landforms

FOUNDN

D

1 The diagram below shows some erosional landforms. Which letter on the diagram relates to the following: **(4 marks)**

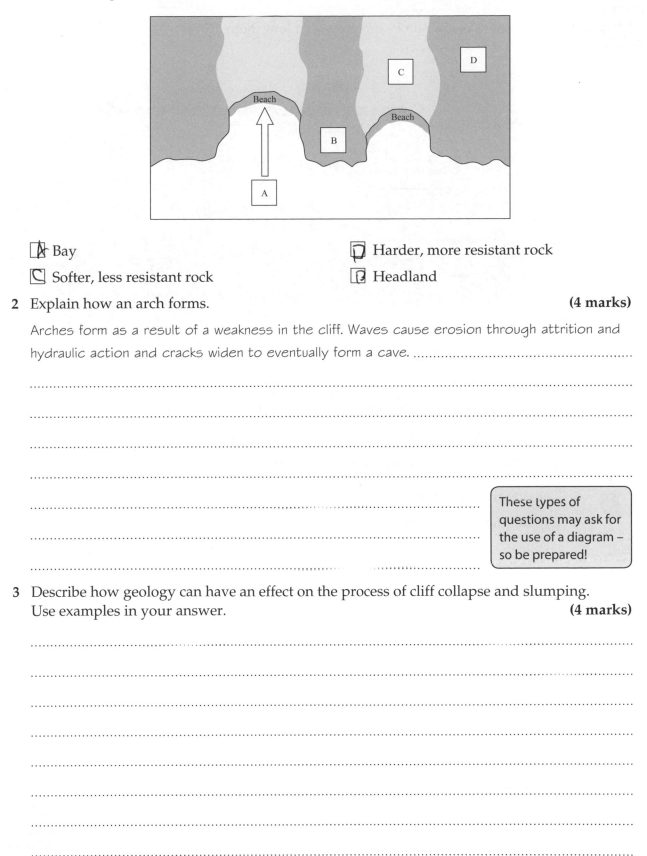

Beach

Beach

C

D

B

A

☒ Bay

☐ Softer, less resistant rock

☐ Harder, more resistant rock

☐ Headland

FOUNDN
HIGHER

C

⟩ Guided ⟩

2 Explain how an arch forms. **(4 marks)**

Arches form as a result of a weakness in the cliff. Waves cause erosion through attrition and

hydraulic action and cracks widen to eventually form a cave. ....

....

....

....

....

....

....

....

....

> These types of
> questions may ask for
> the use of a diagram –
> so be prepared!

HIGHER

B

3 Describe how geology can have an effect on the process of cliff collapse and slumping. Use examples in your answer. **(4 marks)**

....

....

....

....

....

....

....

....

Depositional landforms 1

FOUNDN

D

1 What is the correct term for when wave energy lowers and drops beach sediment? **(1 mark)**

....

FOUNDN

D

2 The diagram below shows the process of longshore drift (LSD).

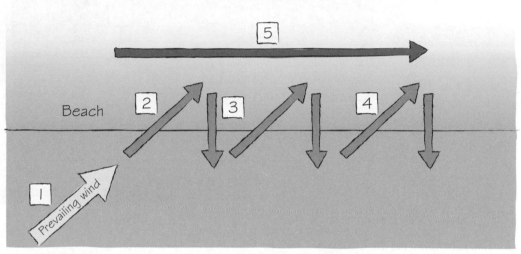

Match the following descriptions to the appropriate label on the diagram. **(5 marks)**

Sediment will move along the beach ☐

The angle of the wind will force LSD to move in the same direction ☐

The backwash will bring the beach sediment down the beach ☐

A 'zig-zag' pattern occurs as the process is repeated ☐

The swash will force the sediment up the beach ☐

HIGHER

B

3 The diagram above shows the process of longshore drift.

Explain the process of longshore drift. **(4 marks)**

Guided

Longshore drift is the movement of beach sediment along a beach. The direction of the wave depends on the prevailing wind ....

....

....

....

....

....

....

> Think about the process as a 'sequence'. This should help you write it up.

Depositional landforms 2

1 The photograph below shows a depositional landform off the coast of Helston, Cornwall.

What are landforms A and B? **(2 marks)**

Choose from the words in the box.

A = ...

B = ...

spit   bar   stack
arch   lagoon

2 The diagram below outlines the geology of the Dorset coastline.

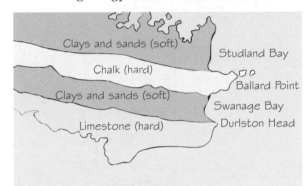

Clays and sands (soft)
Chalk (hard)
Clays and sands (soft)
Limestone (hard)
Studland Bay
Ballard Point
Swanage Bay
Durlston Head

Describe the landforms that are likely to develop. **(6 marks)**

Guided

This deferential geology will cause the formation of headlands and bays. They form because

....

....

....

....

....

....

....

....

....

....

> With any 'formation' question, think about the sequence of formation – this makes it easier to write up!

How coastal landforms change

FOUNDN

D

1 What does the term 'cliff recession' mean? **(1 mark)**

....

....

FOUNDN

C

Guided

2 Explain how differences in the fetch can impact on cliff recession. **(4 marks)**

The longer the fetch, the more energy the wave has. This affects cliff recession because

....

....

Whereas, with a smaller fetch ....

....

....

....

> Fetch is the distance of open water that a wave crosses!

HIGHER

B

3 The digimap shows the rate of cliff recession on the Holderness coast.

Shoreline year
····· 1846
······ 1887
─·─· 1955
─ ─ ─ 1978
──── 1994

0   Kilometres   1

(a) Describe the rate of recession. Use data in your answer. **(3 marks)**

....

....

....

....

....

....

> Start with the earliest year first **and** think about the scale!

(b) Suggest reasons for the rate of cliff recession. **(3 marks)**

....

....

....

....

....

....

Coastal flooding

**FOUNDN
HIGHER
D/C**

Guided

1 Explain how the sea defence in the photo helps
prevent coastal flooding.   **(2 marks)**

The .... helps prevent

coastal flooding by ....

....

....

....

....

**HIGHER
B**

Guided

2 Other than sea defences, describe **one** other method that can be used to minimise the
damage caused by coastal flooding. Use examples in your answer.   **(3 marks)**

One technique that can be used to limit damage from coastal flooding is ....

....

....

....

....

....

> Remember your examples. You
> will limit your marks otherwise!

**HIGHER
A**

3 Explain why the prediction and prevention methods used for coastal flooding differ
between LICs and HICs. Use examples in your answer.   **(6 marks)**

....

....

....

....

....

....

....

....

....

....

....

> You must have a **comparison** between
> your countries in this question.

Coastal protection

FOUNDN
D

1 A sea wall is an example of a hard engineering technique which helps prevent coastal erosion. State **one** advantage and **one** disadvantage of sea walls.   **(2 marks)**

Advantage ....

Disadvantage ....

FOUNDN
C

2 Explain why environmentalists would prefer to use soft engineering rather than hard engineering. Use examples in your answer.   **(4 marks)**

Guided

Environmentalists prefer soft engineering techniques such as ....

because ....

....

Whereas hard engineering techniques such as ....

....

....

FOUNDN
HIGHER
C/B

3 The photograph below is of a hard engineering technique.

(a) What is the name of this technique?   **(1 mark)**

....

(b) How does it help prevent coastal erosion?   **(4 marks)**

Guided

Groynes are used to stop the process of longshore drift. ....

....

....

....

....

....

....

....

> You need to further explain why stopping LSD helps reduce erosion.

Case study

Swanage Bay and Durlston Bay

1   The structure of rocks at the coastline can impact on landforms. When there is a mix of resistant (granite) and less resistant (limestone) rock, which coastal landforms are likely to form?   **(2 marks)**

Landform 1 ....

Landform 2 ....

2   Management of the coast is expensive and so only certain areas in the UK are protected. Using examples you have studied, explain why certain areas of coastline are protected.   **(4 marks)**

....

....

....

....

....

....

....

3   With reference to an example you have studied, outline **two** ways in which the coastline has been managed.   **(4 marks)**

Guided

In Swanage and Durlston Bay, in Dorset, they have used several techniques to protect the coastline. One way in which it is being managed is ....

This helps protect the coastline by ....

....

....

Another way in which it is being protected is through ....

This helps protect the coastline by ....

....   You must know a
specific location.

....

4   In some coastlines, drainage pipes are added to the cliffs. Explain how they prevent cliff collapse.   **(3 marks)**

....

....

....

....

....

Had a go ☐   Nearly there ☐   Nailed it! ☐

Case study # The Cornish coastline

FOUNDN
C

> **Guided**

1 Outline why some coastline areas of the UK are susceptible to high wave attack. **(3 marks)**

The waves that hit the south-west coast of the UK are influenced by the wind direction.

The prevailing wind is from ....

....

....

....

FOUNDN
D

2 State **two** advantages of using soft engineering on the Cornish coastline. **(2 marks)**

1 ....

....

2 ....

....

HIGHER
B

> **Guided**

3 Outline how geology influences the rate of erosion on coasts. Use an example you have studied to help you. **(4 marks)**

The Cornish coastline is made up of very hard geology which is therefore resistant to

....

....

....

....

....

....

....

> You must know a specific location. Lack of appropriate detail will limit the marks.

HIGHER
B

4 Give **three** advantages of dune stabilisation in helping to protect coastlines. **(3 marks)**

....

....

....

....

....

River basins

FOUNDN
E

1 Use words in the text box to complete the following paragraph.   **(5 marks)**

| mouth   streams   uphill   tributary   downhill   watershed |
| upland   confluence   source   west |

All rivers start in .... areas. They can form from bogs, lakes or ....

The point where a river starts is called the .... All rivers flow

down .... and finally end at the .... of the river.

FOUNDN
E

2 The map shows the Amazon drainage basin,
South America. Complete the sentences below
for the features labelled A–C.   **(3 marks)**

▷ **Guided** ▷

Feature A is the s _ _ _ _ _

Feature B is the m _ _ _ _

Feature C is a t _ _ _ _ _ _ _ _

> Make sure you can recognise the
> features of a drainage basin on a
> diagram, map or satellite image.

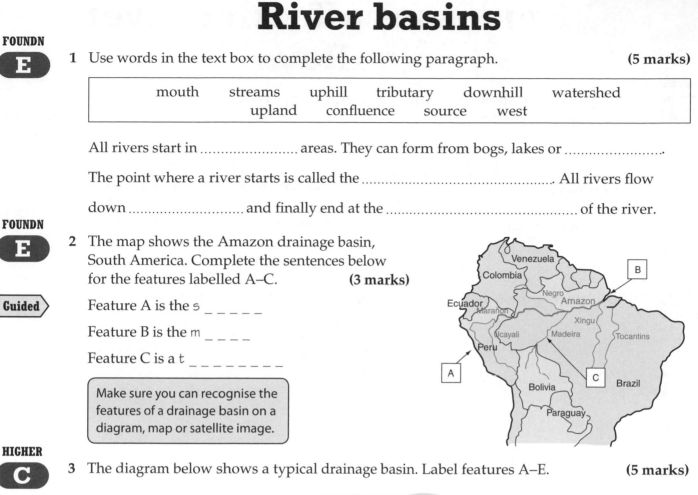

HIGHER
C

3 The diagram below shows a typical drainage basin. Label features A–E.   **(5 marks)**

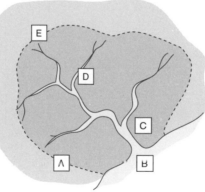

Feature A ....

Feature B ....

Feature C ....

Feature D ....

Feature E ....

Had a go ☐   Nearly there ☐   Nailed it! ☐

Processes affecting river valleys

FOUNDN
D

1 All rivers erode through four main processes. Using label lines, match up the erosion process to the correct definition. **(4 marks)**

Abrasion	Rocks hit against each other within the river water and break up
Corrosion	Air gets squeezed into cracks in the banks and bed, forcing them apart
Hydraulic action	Scraping away of banks and bed by material in the water
Attrition	Chemicals dissolve the minerals in the rock

HIGHER
B

2 Explain the process of freeze–thaw weathering. You may use a diagram(s) to help you.

(4 marks)

....

....

....

....

....

....

....

....

....

....

> Make sure that you use **annotation** rather than labels for your diagrams.

HIGHER
B

3 Compare the processes of abrasion and attrition. **(4 marks)**

> **Guided**

Both abrasion and attrition are forms of erosion but they ....

Abrasion is ....

....

....

Attrition is ....

....

....

....

The long profile

1 The characteristics of a river changes as it moves downstream.

Insert the word **increases** or **decreases** to complete the sentences below. **(3 marks)**

(i) The width of the river .... downstream.

(ii) The gradient of the river .... downstream.

(iii) The velocity of the river .... downstream.

2 The shape of the river channel changes as it moves down the slope.

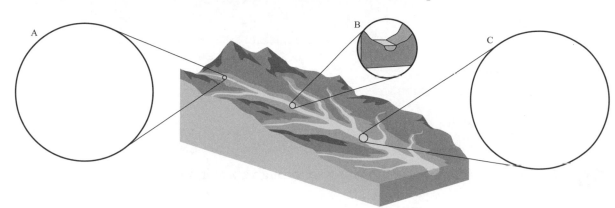

(a) Complete the diagram by sketching the shape of the channel at point A and C. **(2 marks)**

(b) The contour patterns show what the valley looks like at different points on the long profile diagram above. Place the letters A, B and C in the box below each contour pattern to indicate where it would occur on the long profile. **(3 marks)**

(i)    **(ii)**   **(iii)**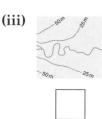

☐   ☐   ☐

(c) Explain your answer for part **(b)** using evidence from the diagrams to help you. **(4 marks)**

The contour lines in diagram **(ii)** are very close together which indicates ....

....

....

....

....

....

....

....

Exam questions similar to
this have proved tricky –
be prepared! ResultsPlus

Upper river landforms

FOUNDN

E

1 The diagram below shows the upper course of a river. What is the feature labelled A called? **(1 mark)**

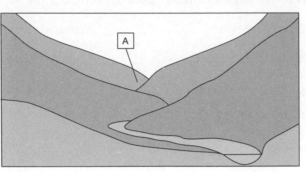

☐ Meander

☐ Slip-off slope

☐ Interlocking spur

☐ Waterfall

HIGHER

B

⟩ Guided ⟩

2 Label the diagram of a cross section through a waterfall. **(3 marks)**

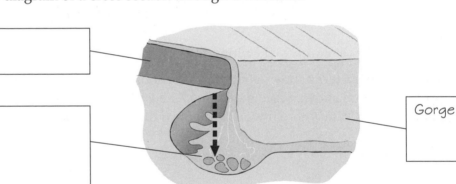

Gorge

HIGHER

B

3 Explain why rock type is important in the formation of a waterfall. Use examples in your answer. **(4 marks)**

....

....

....

....

....

....

....

Include examples of hard and soft rock in your explanation.

Middle river landforms 1

FOUNDN
E

1 Complete the paragraph explaining how a meander is formed using the words from the text box. **(5 marks)**

> slip-off slope   deposition   weathering   fastest   deeper
> bend   channel   erosion   gradient   bed

Read all the text before you start to fill in!

A meander is a .... in a river. They form where the .... of the river is less steep. This means that .... is greatest on the outside bend where the river is .... On the inside bend, the river is slow flowing and so .... occurs.

FOUNDN
HIGHER
C

2 The diagram below shows a cross section through a meander. Label the landforms A and B. **(2 marks)**

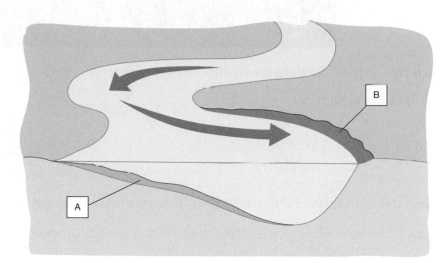

HIGHER
B

A ....   B ....

3 Add the following labels onto the diagram. **(3 marks)**

(i) Slowest flow   **(ii)** Fastest flow   **(iii)** Area of deposition

HIGHER
C

4 What does the term 'lateral erosion' mean? **(1 mark)**

....

....

HIGHER
B

5 Explain how the gradient of a river channel can help the formation of meanders. **(4 marks)**

In the middle course of the river the gradient has become ....

Guided

....

....

....

....

....

....

Think about the features of the long profile of the river!

Middle river landforms 2

FOUNDN

D

1 The aerial photograph shows the Danube Delta in Romania. What is the name of river landform A? **(1 mark)**

Landform A =

FOUNDN

D

2 Which **two** of the following are middle river course landforms? **(2 marks)**

☐ Interlocking spur

☐ River cliff

☐ Waterfall

☐ Meander

> Look through all the options first, and think about where they formed on the river!

HIGHER

B

⟩ **Guided** ⟩

3 Explain the formation of an ox-bow lake. You may use a diagram to help. **(4 marks)**

Ox-bow lakes form when rivers create meanders. Deposition occurs ....

....

....

....

....

....

....

....

Lower river landforms

FOUNDN
D

1 The aerial image shows the Zambezi River, Africa. Name the river landforms labelled A and B.

(2 marks)

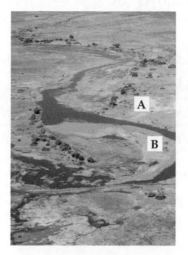

A ....

B ....

FOUNDN
C

HIGHER
B

⟩ Guided ⟩

2 Explain why deposition occurs in the lower course of the river.   **(4 marks)**

The gradient of the river is at its lowest in the lower course of the river. This ....

....

....

....

....

....

....

....

> Make sure you comment on the size of the alluvium that gets deposited first **and** why!

HIGHER
B

3 Explain the formation of a natural levee. You may use diagrams to help.   **(4 marks)**

....

....

....

....

....

....

....

....

....

Why do rivers flood?

FOUNDN
E

1 Are the following statements about flooding **true** or **false**?

Put a tick (✓) in the correct box.   **(4 marks)**

True	False	
		Flooding is when a river reaches the top of its bank and spills over
		Flooding occurs on all rivers
		Floods can occur because there is too much rain
		Afforestation can cause flooding

FOUNDN
D

2 Read through the newspaper clippings below.

Shade in red all the physical causes of flooding. Shade in blue the human causes of flooding.   **(5 marks)**

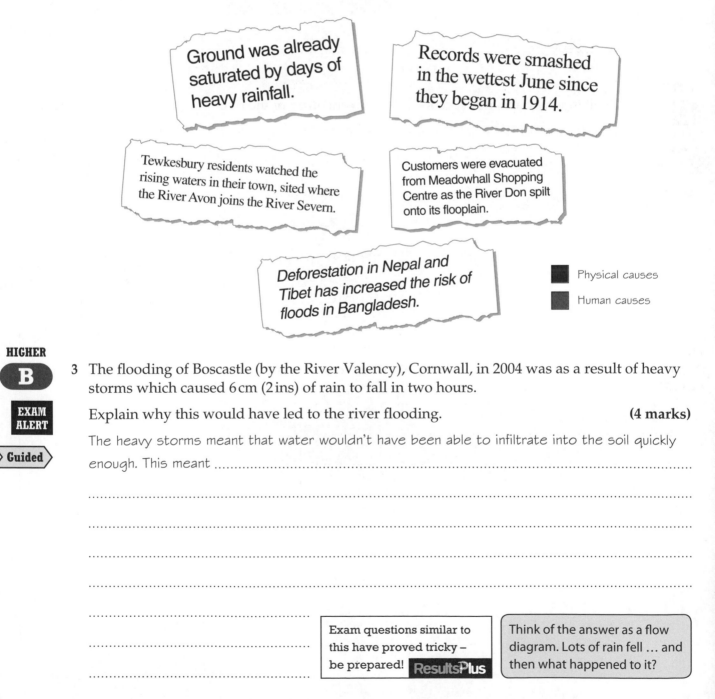

Ground was already saturated by days of heavy rainfall.

Records were smashed in the wettest June since they began in 1914.

Tewkesbury residents watched the rising waters in their town, sited where the River Avon joins the River Severn.

Customers were evacuated from Meadowhall Shopping Centre as the River Don spilt onto its flooplain.

Deforestation in Nepal and Tibet has increased the risk of floods in Bangladesh.

■ Physical causes

■ Human causes

HIGHER
B

EXAM ALERT

▷ Guided ▷

3 The flooding of Boscastle (by the River Valency), Cornwall, in 2004 was as a result of heavy storms which caused 6 cm (2 ins) of rain to fall in two hours.

Explain why this would have led to the river flooding.   **(4 marks)**

The heavy storms meant that water wouldn't have been able to infiltrate into the soil quickly

enough. This meant ....

....

....

....

....

....

| Exam questions similar to this have proved tricky – be prepared! **ResultsPlus** | Think of the answer as a flow diagram. Lots of rain fell … and then what happened to it? |

Case study

River Stour, Dorset

FOUNDN
D

1   Some rivers need to be managed to prevent flooding. One way of doing this is through hard engineering techniques such as flood walls.

Give **one** advantage and **one** disadvantage of using flood walls.   **(2 marks)**

Advantage ....

....

Disadvantage ....

....

HIGHER
B

2   With reference to an example you have studied, explain why a change in land use can cause rivers to flood.   **(4 marks)**

Guided

The deforestation of woodland would mean that there are fewer trees and leaves to

intercept the rain water, therefore reducing interception storage. ....

....

....

....

....

.... | Try to include correct river terminology in your explanations. |

HIGHER
A

3   Using an example you have studied, explain the physical and human causes of the flooding.   **(6 marks)**

....

....

....

....

....

....

....

....

....

....

....

| Aim to include relevant examples and detail to support your points and make sure that your answer is logical and well-organised. |

Had a go ☐   Nearly there ☐   Nailed it! ☐

Case study Mississippi, USA (2011)

FOUNDN
F

1 Use some of the words in the box to complete the sentences below.   **(5 marks)**

> water   land   human   rainfall
> environmental   river·   impermeable
> slower   permeable   quicker

Cross through the words in the text box once used!

Rivers flood for many different reasons. These can be linked to physical and / or .... reasons.

Most rivers flood because there is lots of ...., which has created excess water.

Another factor is when humans build on the flood plain. This is called .... use change. Building on the flood plain increases the number of .... (water cannot pass through) surfaces and so water run-off is much ....

FOUNDN
HIGHER
C/B
Guided

2 Using an example of a flood you have studied, describe the effects of the flood.   **(4 marks)**

The flooding of the Mississippi in 2011 caused massive impacts. There were impacts on the people, environment and the economy. ....

....

....

....

....

....

....

Make sure you have case study detail in this type of question.

HIGHER
B

3 With reference to an example of a flood event you have studied, explain **two** ways in which the flood was managed.   **(4 marks)**

....

....

....

....

....

....

....

....

Reducing the impact of flooding

FOUNDN
C

1 (a) This symbol is commonly used by the Environment Agency. What does it mean? **(1 mark)**

....

(b) Why is it important to educate individuals about flooding? **(3 marks)**

....

....

....

....

....

....

HIGHER
B

2 The map shows the areas of London that are at risk from flooding.

Key
Areas at risk of flooding

Islington Havering Barking and Dagenham Newham Tower Hamlets Westminster City Thames Barrier Kensington and Chelsea Southwark Greenwich Battersea Lewisham Bexley Richmond upon Thames Wandsworth Lambeth Merton

5 km

You will need to describe at least **two** patterns or trends. 1 mark will be allocated for evidence.

Describe the flood risk pattern. Use evidence from the map to help. **(3 marks)**

....

....

....

....

....

HIGHER
B

3 Describe how prediction of a flood event can help reduce the potential impacts from flooding. **(4 marks)**

Guided

The Environment Agency monitors rivers and gains information on weather patterns from the Met Office in the UK. They use this information to predict ....

....

....

....

....

....

Flood management: soft engineering

FOUNDN
D

1 Give **one** advantage and **one** disadvantage of using soft engineering techniques.   **(2 marks)**

Advantage ....

....

Disadvantage ....

....

HIGHER
B

2 Planting trees in the catchment area of a river can (as shown in the photograph opposite) help prevent flooding. Explain why.
   (4 marks)

....

....

....

....

....

....

....

....

....

....

> Try to remember the correct terminology for soft engineering techniques, e.g. afforestation.

HIGHER
B

3 Explain why some groups of people support and other groups oppose soft engineering techniques to reduce flooding.
   (4 marks)

⟩ Guided ⟩

Soft engineering is an environmentally friendly form of managing flooding, however many farmers would prefer not to use storage areas because ....

....

....

....

....

....

....

Flood management: hard engineering

FOUNDN
E

1 What does the term 'hard engineering' mean? **(1 mark)**

....

....

FOUNDN
D

2 Give **two** examples of hard engineering that can be used to prevent river flooding. **(2 marks)**

Technique 1 ....

Technique 2 ....

FOUNDN
HIGHER
C/B

**EXAM
ALERT**

3 (a) This photo is an example of a flood prevention technique on a river in the USA.

Exam questions similar to this have proved tricky – be prepared! **ResultsPlus**

(i) What is the engineering technique called? **(1 mark)**

....

(ii) Describe how it can prevent flooding. **(2 marks)**

....

....

....

....

(b) Explain why some people would be opposed to using hard engineering techniques. Use examples in your answer. **(4 marks)**

Guided

When the Three Gorges Dam was built in China, many local residents were against it

because ....

....

....

....

....

....

....

Try to refer to **named** groups of people rather than simply stating 'people'. Include more than one example.

Earthquakes and volcanoes

FOUNDN
E

1 The map here shows the global locations of volcanoes and earthquakes.

Are the following statements about the global locations of volcanoes and earthquakes true or false? **(5 marks)**

> Think about the **general** pattern across the whole map. Don't just focus on a particular country or region.

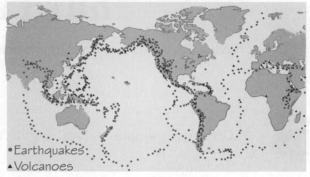

• Earthquakes
▲ Volcanoes

True	False	
		All volcanoes and earthquakes occur on the coast
		Volcanoes and earthquakes occur at plate boundaries
		Some volcanoes occur in the middle of plates
		There are no volcanoes in South America
		Volcanoes and earthquakes occur in narrow bands

HIGHER
C

2 What is the correct term for volcanoes which form in the middle of tectonic plates? **(1 mark)**

....

HIGHER
C

3 The map here shows the magnitude and focal depth of earthquakes in New Zealand since 1843.

Describe the distribution of the earthquakes shown by the map.
(4 marks)

> **Guided**

The largest earthquakes (>mag 7) are located where the focal depth is the smallest, e.g. south of the island

....

....

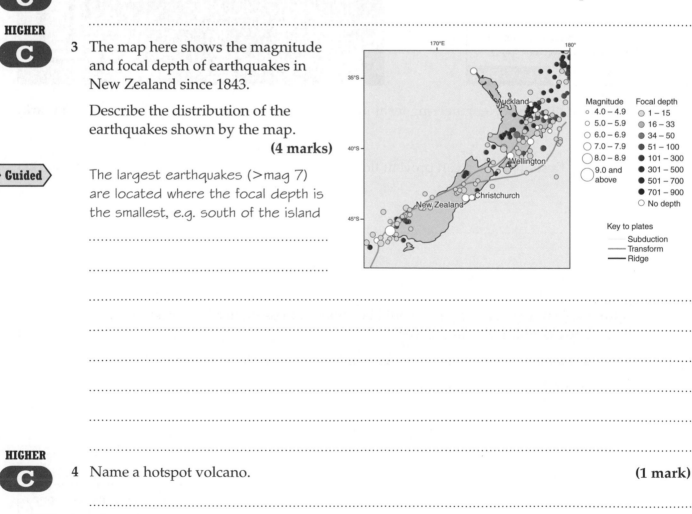

....

....

....

....

....

....

HIGHER
C

4 Name a hotspot volcano. **(1 mark)**

....

Plate tectonics

FOUNDN
D

1 Draw lines to match the plate movement to the name of the plate boundary.   **(3 marks)**

Divergent (constructive)

Convergent (destructive)

Conservative (transform)

HIGHER
B

2 With the aid of a diagram, explain the process by which plates move.   **(4 marks)**

....

....

....

....

....

....

....

....

....

....

HIGHER
B

Guided

3 Explain how the island chain of Hawaii formed.   **(4 marks)**

Hawaii is a series of volcanic islands which formed because of a hotspot in the Earth's crust.

A hotspot means that the crust is thinner or weaker which means that ....

....

....

....

....

....

....

....

Remember you need to
explain the formation
of a hotspot volcano!

Convergent plate boundaries

FOUNDN
C
Guided

1 How does a convergent plate margin differ from a conservative plate margin?   **(3 marks)**

Conservative plate margins move side by side whereas convergent plate margins ....

....

....

....

....

....

> Try not to include the similarities between the plate margins. The question has specified the differences.

HIGHER
B

2 Some convergent plate boundaries have both oceanic and continental crust.

State **three** differences between oceanic and continental crust.   **(3 marks)**

....

....

....

....

....

....

HIGHER
B

3 Explain why violent earthquakes and volcanoes occur at convergent plate margins.

You may use a diagram to help you.   **(4 marks)**

....

....

....

....

....

....

....

....

....

....

Divergent plate boundaries

FOUNDN

E

1 Are the following statements about divergent plate margins true or false? **(4 marks)**

T	F	
		Divergent plates move together.
		Only earthquakes occur at divergent plates.
		Volcanoes form and earthquakes occur at divergent plates.
		Sea floor spreading occurs at divergent plates.

FOUNDN

D

2 The following statements show the formation of volcanoes in the Atlantic Ocean. Number the statements in order of formation of the volcanic island. **(6 marks)**

> **Guided**

Order of formation	
	The magma cools and solidifies.
1	The North American plate and Eurasian plate move apart.
	Over time, the ridge gets bigger in size.
	Magma rises through weaknesses in the crust.
	This forms a ridge on the sea floor.
	Eventually, the volcano will appear above the surface of the sea.

HIGHER

B

3 Describe how volcanoes form at divergent plate margins. You may use a diagram to help you. **(4 marks)**

> **Guided**

At a divergent plate margin, the plates move

....

....

....

....

....

....

....

....

....

> To do well, aim to include a fully annotated (explanatory) diagram.

Conservative (transform) plate boundaries

1 Complete the sentences below. Use the words in the box below to help.   **(5 marks)**

| together   volcanoes   stuck   subducted   side by side   loose |
| volcano   heat   pressure   earthquakes   earthquake |

In a conservative plate margin the plates move .... None of the crust

is being destroyed and so only .... occur. Sometimes, the plates can get

.... This can increase .... which is suddenly

released, causing an .... An example of this plate margin is the

San Andreas Fault.

2 What is another name for a conservative plate margin?   **(1 mark)**

....

3 The aerial photograph below is of the San Andreas Fault in California.

Name plates A and B   **(2 marks)**

Plate A ....

Plate B ....

4 Outline why volcanoes do not occur along conservative plate boundaries.   **(2 marks)**

....

....

5 How do the characteristic features of a divergent plate margin differ from a
conservative?   **(4 marks)**

In a conservative plate margin, plates move side by side ....

....

....

....

....

....

....

....

....

In this type of question, try to
use the words '**compared** to' as
this will help you highlight to
the examiner the differences
between the two plate margins.

Exam questions similar to
this have proved tricky –
be prepared! **Results**Plus

Measuring earthquakes

FOUNDN
D

1 What is the 'focus' of an earthquake? **(1 mark)**

....

....

FOUNDN
D

2 Name features A and B
on the diagram of an
earthquake below. **(2 marks)**

B

Fault

A

A ....

B ....

HIGHER
B

3 What **two** scales are used to measure earthquakes? **(2 marks)**

....

....

HIGHER
C

4 Give **three** differences between the two scales. **(3 marks)**

⟩ **Guided** ⟩

.... measures ....

.... measures ....

.... has no upper limit. .... has a limit of 12.

....

....

....

HIGHER
C

5 (i) The image depicts the instrument used
to measure earthquakes. What is it
called? **(1 mark)**

....

....

(ii) Outline how the instrument in the image
measures earthquakes. **(3 marks)**

....

....

....

....

....

Living with hazards

FOUNDN
D

1 The image shows one of the benefits of living in areas close to tectonic activity.

What is the benefit?   **(1 mark)**

....

FOUNDN
C

EXAM ALERT

2 Give **two** other benefits of living in areas of tectonic activity. Include a located example for each.   **(4 marks)**

Benefit 1 ....

....

....

Benefit 2 ....

....

....

> Exam questions similar to this have proved tricky – be prepared! **ResultsPlus**

HIGHER
B

3 The map below shows the density of the population around Mt Tambora in the Asia-Pacific Region.

(a) Describe the population density. Use data in your answer.   **(4 marks)**

Guided

It is clear to see that the population density is unevenly distributed. ....

....

....

....

....

....

....

....

(b) Explain why people would still choose to live close to the volcano.   **(3 marks)**

....

....

....

Case study **Chances Peak volcano**

FOUNDN
D

1 Complete the sentences below. Use the words in the box below to help.   **(5 marks)**

lava	Montserrat	Lanzarote	1985	January	pyroclastic
	June	extinct	1995	dormant	

Chances Peak is an active volcano on the island of .... in the Caribbean.
Prior to eruption, the volcano was .... but became active in
.... The major eruption occurred on 25 .... 1997.
The main volcanic hazard was from a .... flow.

FOUNDN
C

2 Describe the environmental impacts of a volcanic eruption you have studied.   **(4 marks)**

....

....

....

....

....

....

....

....

> The point of the question is impact on the **environment**. Do not include impacts on people or the economy.

HIGHER
A

3 With reference to a volcanic eruption you have studied, explain why the volcano erupted.   **(4 marks)**

Guided

Chances Peak in Montserrat is located on a convergent plate margin. This is where two

plates are moving together. ....

....

....

....

....

....

....

HIGHER
B

4 With the aid of an example, outline the social impacts of a volcanic eruption.   **(3 marks)**

....

....

....

....

....

....

Case study Bam earthquake (2003)

FOUNDN
C

1 The time that an earthquake occurs can affect the death toll.

Outline why an earthquake which happens very early in the morning would be more likely to have a higher death toll than one during the middle of the day.   **(2 marks)**

....

....

....

....

HIGHER
A

2 Suggest possible impacts of an earthquake on the environment. Use case study detail in your answer.   **(4 marks)**

....

....

....

....

....

....

....

....

> In the exam you might be asked about either an earthquake or a volcanic eruption and its effects on people and the environment.

HIGHER
A

Guided

3 Explain how the impacts of an earthquake can be dealt with. Use case study detail in your answer.   **(6 marks)**

The Bam earthquake in Iran in 2003, initiated immediate short-term responses, but they also had long-term responses as well. ....

....

....

....

....

....

....

....

....

....

> If possible make the distinction between short and long-term effects or responses.

Preventing tectonic hazards

FOUNDN

D

1 The photograph below shows the Transamerican building in San Franscisco.

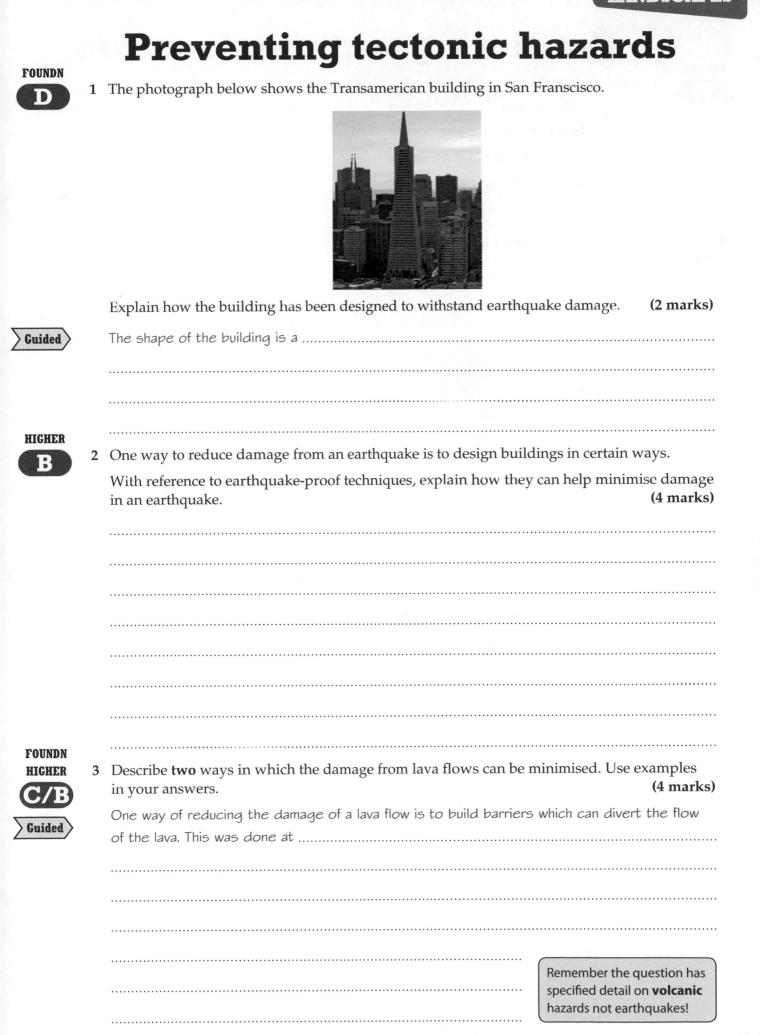

Explain how the building has been designed to withstand earthquake damage. **(2 marks)**

Guided

The shape of the building is a ....

....

....

....

HIGHER

B

2 One way to reduce damage from an earthquake is to design buildings in certain ways.

With reference to earthquake-proof techniques, explain how they can help minimise damage in an earthquake. **(4 marks)**

....

....

....

....

....

....

....

....

FOUNDN

HIGHER

C/B

3 Describe **two** ways in which the damage from lava flows can be minimised. Use examples in your answers. **(4 marks)**

Guided

One way of reducing the damage of a lava flow is to build barriers which can divert the flow

of the lava. This was done at ....

....

....

....

....

....

Remember the question has specified detail on **volcanic** hazards not earthquakes!

Had a go ☐   Nearly there ☐   Nailed it! ☐

Predicting tectonic hazards

FOUNDN
F

1 There are many instruments used to predict when tectonic hazards will occur. What does a seismometer measure? **(1 mark)**

☐ **A** The amount of sulphur dioxide given off

☐ **B** The temperature of the magma

☐ **C** The shaking of the ground (earthquakes)

☐ **D** The ash emitted by the volcano

FOUNDN
HIGHER
C/B

2 Describe **two** methods used to predict when a volcano is likely to erupt. **(4 marks)**

Method 1 ....

....

....

....

Method 2 ....

....

....

....

> It is important that you know which **instruments** can be used to help predict volcanic eruptions.

HIGHER
B
Guided

3 Explain how satellite images could be used to predict a potential volcanic eruption. **(3 marks)**

Satellite images use infrared which measures ....

....

....

....

....

....

....

> Remember that a satellite is a camera in space. It is not an instrument used on the Earth's surface.

A world of waste

FOUNDN
D

1 (a) The bar graph below shows the amount of waste generated per person per day for a number of cities. Complete the graph using the data from the table. **(2 marks)**

Dhaka, Bangladesh	0.1
Paris, France	1.3

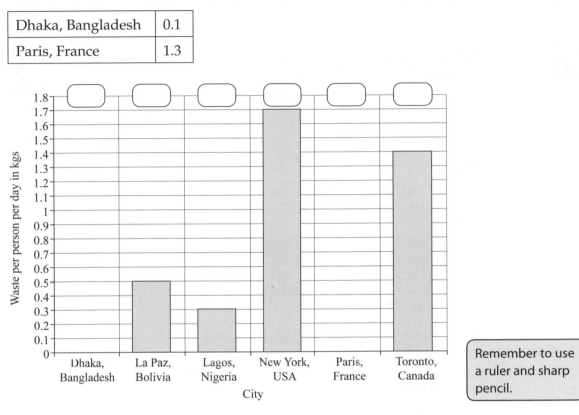

Waste per person per day in kgs

Cities (x-axis): Dhaka, Bangladesh · La Paz, Bolivia · Lagos, Nigeria · New York, USA · Paris, France · Toronto, Canada

City

Remember to use a ruler and sharp pencil.

(b) In the boxes above each bar graph, state whether each example is from a LIC (**L**) or HIC (**H**). **(3 marks)**

HIGHER
C

2 HICs commonly dispose of e-waste (electronic waste).

State **two** examples of e-waste. **(2 marks)**

Example 1 ....

Example 2 ....

HIGHER
C

3 Compare the waste generated in a HIC with a typical LIC. Use examples in your answer. **(4 marks)**

....

....

....

....

....

....

....

....

Compare means describing the similarities and differences!

Wealth and waste

FOUNDN

D

1 What does the term 'throw-away society' mean? Include an example.　　　**(2 marks)**

....

....

HIGHER

B

2 (a) The scatter graph shows the comparison between the wealth and the amount of waste
generated for different countries.

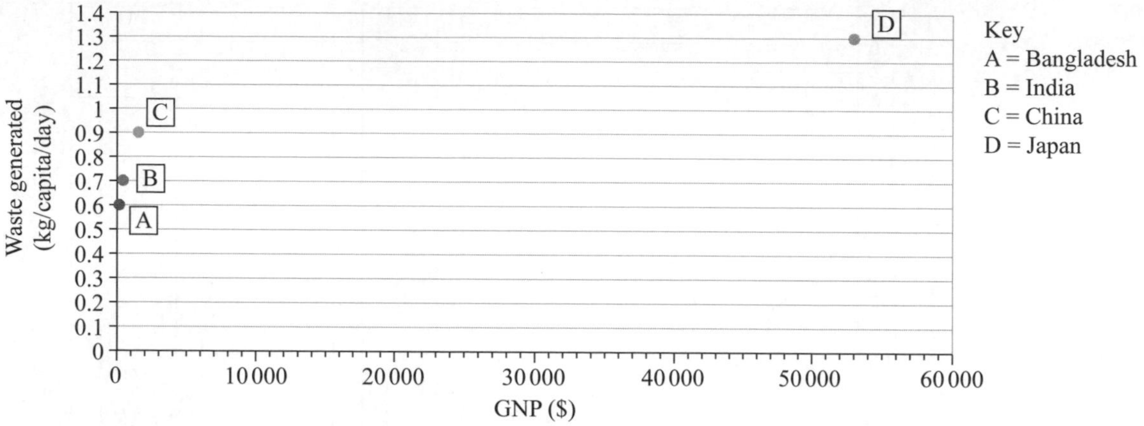

Key
A = Bangladesh
B = India
C = China
D = Japan

(i) Complete the graph using the data in the table below – label it 'Country E'.　**(2 marks)**

Country E	**GNP**	**Waste generated**
	36 000	1.1

(ii) Describe the pattern shown by the completed graph.　　　**(2 marks)**

....

(b) Explain why Bangladesh (country A) generates a low amount of waste.　**(4 marks)**

Guided

Bangladesh is a very low income country with a low GDP, which means ....

....

....

....

....

....

....

Domestic waste in HICs

FOUNDN
E

1 The list below shows some of the different types of waste which can be produced in a country.

Tick (✓) all the waste that can be classified as household. **(2 marks)**

☐ Nuclear   ☐ Toxic   ☐ Industrial

☐ Kitchen scraps ☐ Plastic

☐ Cardboard   ☐ Aluminium cans

> Even though the question is worth 2 marks, there may be more than 2 answers! Read all of them.

FOUNDN
HIGHER
C

2 The choropleth map opposite shows the amount of electronic waste (e-waste) collected from private households in Europe in 2008.

Describe the patterns shown by the map. **(4 marks)**

Guided

The pattern of e-waste collection is
not evenly distributed. ....

....

....

....

....

....

....

....

WEEE collection rate, 2010 (kg per capita)

⬚ <= 1.0
⬚ 1.0–4.0
⬚ 4.0–8.0
■ >8.0
▦ Data not available

Sweden 15.9
Finland 9.1
Norway 15.8
Estonia 4.2
Latvia 1.9
Denmark 14.8
Lithuania 2.7
Ireland 8.2
UK 7.4
Netherlands 7.3
Poland 2.8
Germany 8.8
Czech Republic 5.0
Slovakia 3.9
Belgium 9.3
Luxembourg 0.4
Austria 8.7
Hungary 3.9
Romania 1.1
France 6.4
Liechtenstein 1.4
Italy 4.2
Bulgaria 5.9
Portugal 4.4
Spain 3.2
Greece 3.9
Malta 3.4
Cyprus 3.2

Data source: Eurostat Administrative boundaries:
© EuroGeographics © UN-FAO © Turkstat
Cartography: Eurostat – GISCO, 10/2012

> Include data to support your answer.

HIGHER
B

3 Explain why the **type** of household waste differs between HICs and LICs. **(4 marks)**

....

....

....

....

....

....

....

....

> You do not need to go through **every type** of household waste!

Recycling waste locally

FOUNDN
D

1 Is recycling or reducing waste the most sustainable?

Suggest reasons for your answer.   **(3 marks)**

> **Guided**

The most sustainable is .... because ....

....

....

....

....

....

FOUNDN
D

2 Give **three** advantages of recycling waste.   **(3 marks)**

1 ....

....

2 ....

....

3 ....

....

HIGHER
C

3 Another way of disposing of waste is through landfill. What is landfill?   **(2 marks)**

....

....

HIGHER
C

4 State **one** advantage and **one** disadvantage of incineration as a method of waste disposal.   **(2 marks)**

Advantage ....

....

Disadvantage ....

....

Recycling waste locally: Camden

FOUNDN
E

1 A method of managing waste in a more sustainable way is to implement the 3 Rs. Recycling is one. Name the other two. **(2 marks)**

R............................ R...........................

FOUNDN
D

2 Describe what happens to recycled glass in an example you have studied. **(3 marks)**

....

....

....

....

....

....

FOUNDN
HIGHER
C

3 The line graph below shows Camden's household waste recycling rates between 2004 and 2010.

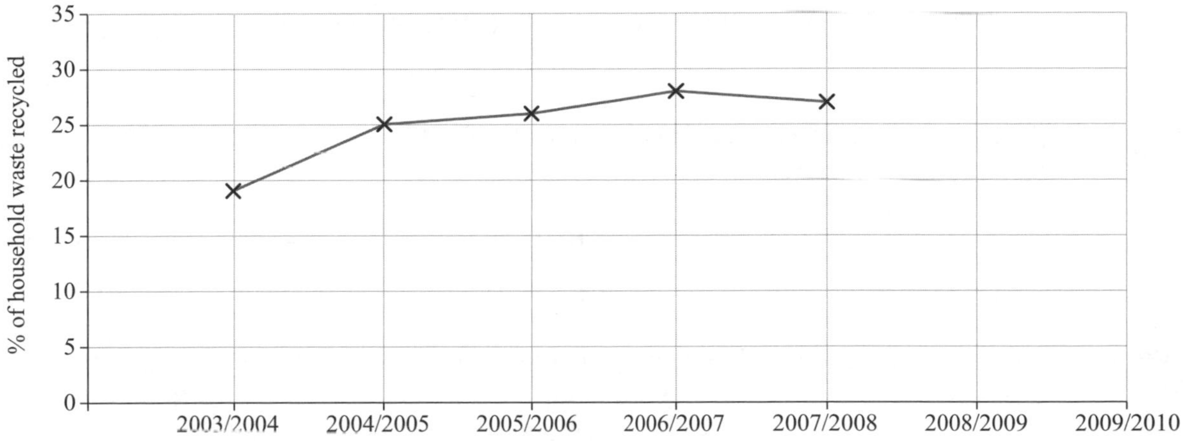

(a) Complete the graph using the data in the table below. **(2 marks)**

2008 09	30%
2009–10	35%

(b) How much have recycling rates increased between 2003 and 2010? **(1 mark)**

....

HIGHER
B

Guided

4 Outline the strategies used to encourage recycling. **(4 marks)**

Camden in North London has achieved a recycling rate of 27.2% in 2007–08 (I). To help

increase these this it ....

....

....

....

....

| An example will always help in questions such as these! |

Waste disposal in HICs

FOUNDN
D

1 (a) The graph below shows both the amount of packaging produced and the amount of recycling of packaging per person in some European countries.

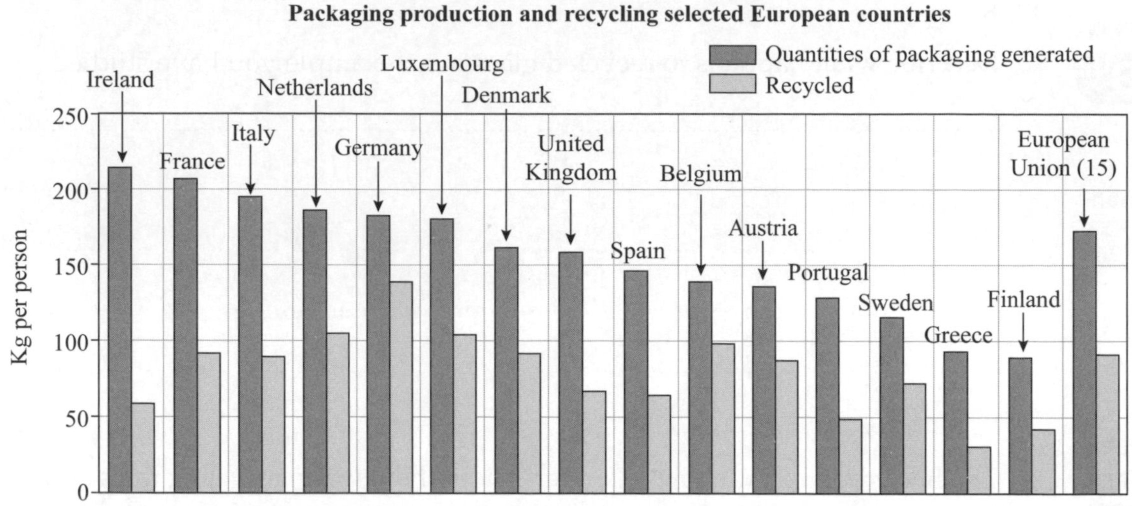

Packaging production and recycling selected European countries

■ Quantities of packaging generated
□ Recycled

Which country has the lowest recycling rate per person? **(1 mark)**

....

FOUNDN
E

2 Which of the countries listed below recycle less than 50% of the overall packaging waste generated? **(3 marks)**

☐ **A** Ireland　　☐ **B** Belgium　　☐ **C** Germany

☐ **D** France　　☐ **E** Austria　　☐ **F** Portugal

FOUNDN
C

HIGHER
A*

Guided

3 Using an example you have studied, explain how successful recycling can be achieved. **(6 marks + 4 marks SPaG)**

Germany recycles 60% of all its household waste ....

....

....

....

....

....

....

....

....

....

....

....

....

> Include relevant case study specific detail here, and leave time to check your spelling, punctuation and grammar.

Case study # Non-renewable energy

FOUNDN
E

1 Use some of the words in the box to complete the sentences below. **(5 marks)**

| infinite | solar | once | finite | fossil fuels | coal | wind | oil |

Non-renewable resources are ones that can only be used ....

They often come from .... These energy resources are often

...., such as .... and ....

FOUNDN
C

2 State **three** advantages of the extraction of a non-renewable energy source for a named country you have studied. **(3 marks)**

Named country ....

1 ....

....

2 ....

....

3 ....

....

HIGHER
A*

3 Describe the consequences of the extraction of a non-renewable source of energy. Use an example in your answer. **(6 marks + 4 marks SPaG)**

Chosen example ....

⟩ **Guided** ⟩ There have been many consequences of drilling for oil in .... There are

positive impacts as well as negative. ....

....

....

....

....

....

....

....

....

....

....

Remember that the word
'consequences' can refer
to the **costs** and **benefits**.

Had a go ☐ Nearly there ☐ Nailed it! ☐

Case study Renewable energy

FOUNDN D

1 Which of the following energy sources can be classified as renewable? **(3 marks)**

☐ **A** Wind ☐ **D** Solar

☐ **B** Oil ☐ **E** Natural gas

☐ **C** Nuclear power ☐ **F** Geothermal

Remember that renewable energy can be used more than once!

FOUNDN E

2 State **one** advantage and **one** disadvantage of using wind as a source of power. **(2 marks)**

Advantage ....

Disadvantage ....

HIGHER C

3 The map shows the positions of both onshore and offshore wind farms in the UK.

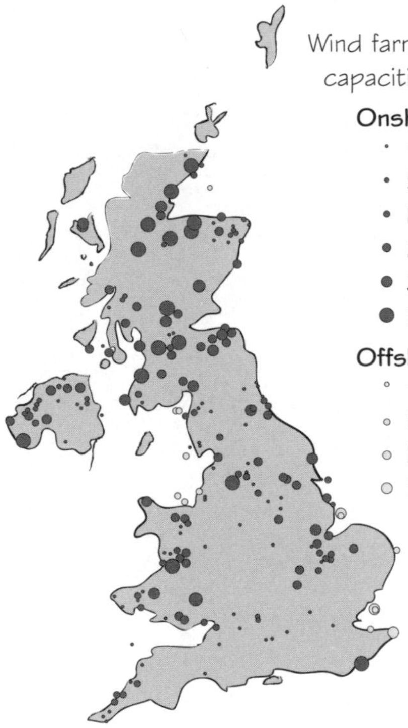

Wind farm installed capacities (MW)
Onshore
< 5
5 – 10
10 – 20
20 – 30
30 – 40
> 40
Offshore
< 10
10 – 100
100 – 500
> 500

Describe the distribution of onshore wind farms. Use evidence from the map in your answer. **(4 marks)**

Guided

From the map, it is clear to see that the majority of onshore turbines are located in Scotland.

....

....

....

....

....

....

....

....

Do not describe **all** the turbines. The question is looking at the **general patterns**.

Energy deficit and surplus

FOUNDN
D

1 What is energy surplus? Put a tick (✓) next to the correct answer. **(1 mark)**

☐ More energy is used than produce   ☐ All the energy produced is used

☐ All the energy produced is sold   ☐ More energy is produced than use

FOUNDN
E

2 The map shows countries which have an energy surplus. Use the map to decide if the following statements are **true** or **false**. **(5 marks)**

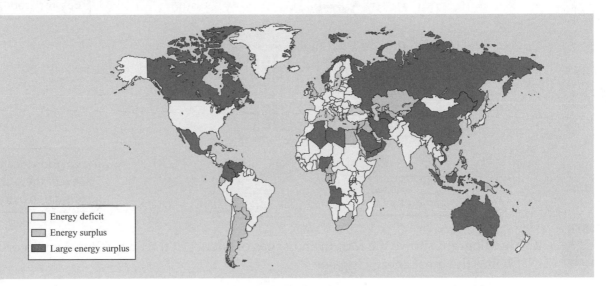

True	False	
		The distribution of energy surplus is evenly distributed
		There are more HICs with a surplus than LICs
		Australia has an energy surplus
		Most of Europe has an energy surplus
		The UK has an energy surplus

HIGHER
C

3 What does it mean when a country has an energy balance? **(1 mark)**

....

....

HIGHER
B

Guided

4 Explain why countries such as Germany currently have an energy deficit. **(4 marks)**

Germany has an energy deficit because it is an HIC but has limited sources of coal and oil.

This means ....

....

....

....

....

....

In this question, you need to think about **sources** of energy and what they need it for.

Wasting our energy

FOUNDN
D

EXAM ALERT

1 The diagram below shows how heat is lost from a typical house.

Complete the diagram by adding the following percentages into the correct spaces: – 45%, 35%, 10%.

(3 marks)

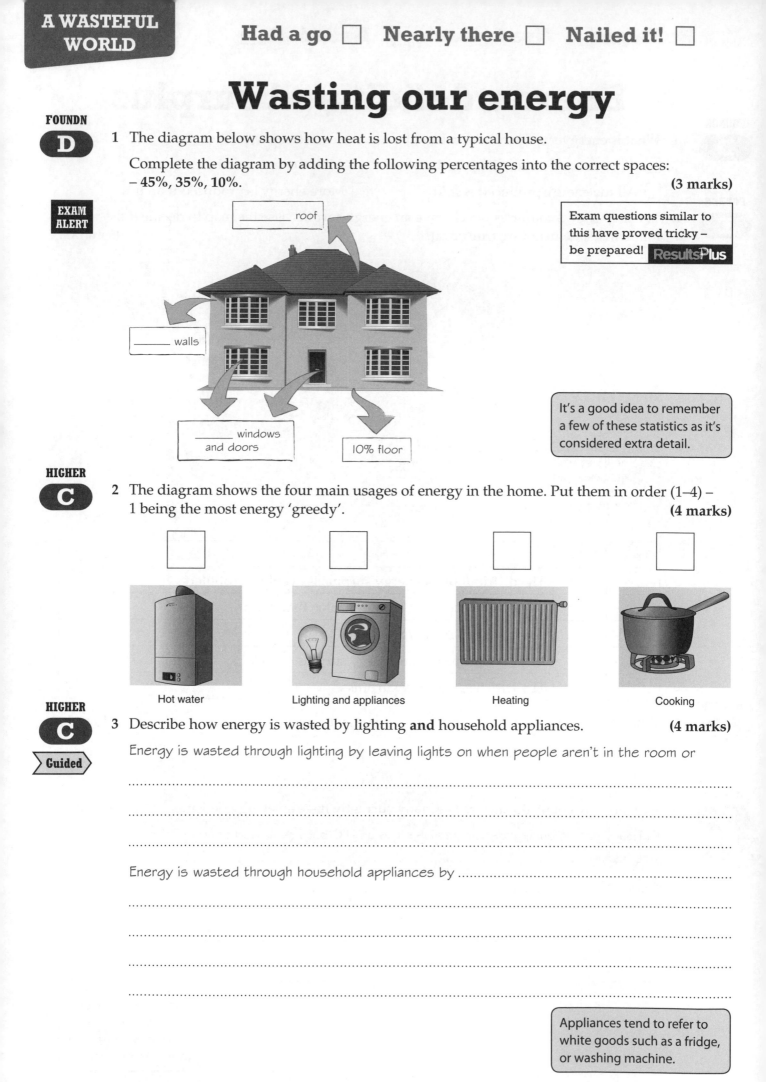

_____ roof

_____ walls

_____ windows and doors

10% floor

Exam questions similar to this have proved tricky – be prepared! ResultsPlus

It's a good idea to remember a few of these statistics as it's considered extra detail.

HIGHER
C

2 The diagram shows the four main usages of energy in the home. Put them in order (1–4) – 1 being the most energy 'greedy'.

(4 marks)

☐ ☐ ☐ ☐

Hot water   Lighting and appliances   Heating   Cooking

HIGHER
C

Guided

3 Describe how energy is wasted by lighting **and** household appliances. **(4 marks)**

Energy is wasted through lighting by leaving lights on when people aren't in the room or

....

....

....

Energy is wasted through household appliances by ....

....

....

....

....

Appliances tend to refer to white goods such as a fridge, or washing machine.

Carbon footprints

FOUNDN
D

1 What does the term 'carbon footprint' mean? **(2 marks)**

....

....

| Remember that it's not just carbon dioxide emissions! |

....

....

FOUNDN
D

2 State **three** activities that contribute towards an individual's carbon footprint. **(3 marks)**

1 ....

2 ....

3 ....

FOUNDN
HIGHER
C

3 (a) The graph below shows the carbon footprint per person of six different countries.

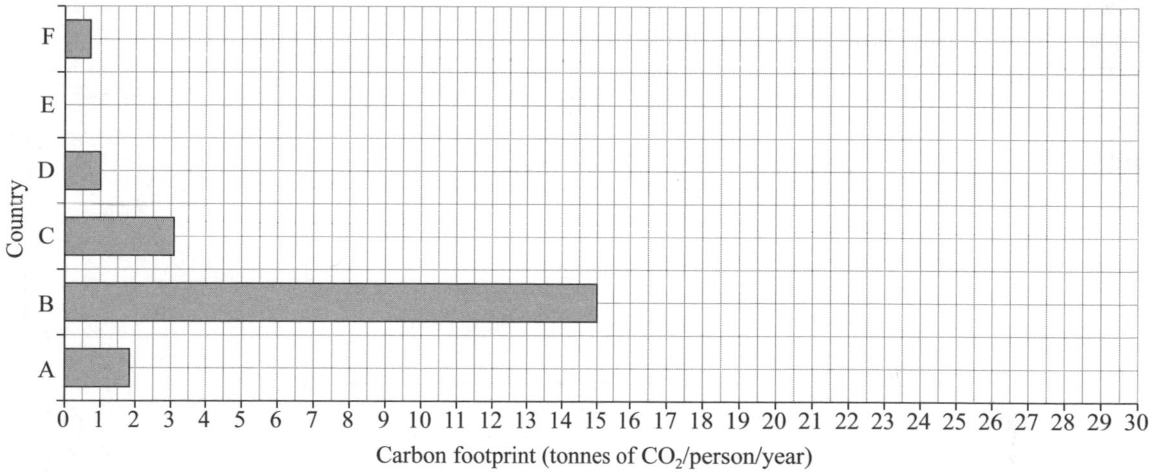

Carbon footprint (tonnes of CO_2/person/year)

Complete the graph by plotting the value of 29 for country E. **(1 mark)**

(b) (i) Which **two** countries are likely to be HICs? **(2 marks)**

Country .... and country ....

(ii) Which two countries are likely to be LICs? **(2 marks)**

Country .... and country ....

HIGHER
B

4 As a country becomes more developed its carbon footprint tends to increase. Suggest reasons why. **(3 marks)**

Guided

As a country becomes more developed the people who live there have more disposable

income so ....

....

....

....

....

....

Energy efficiency

FOUNDN
E

1 What does the term 'energy efficiency' mean? **(1 mark)**

....

....

FOUNDN
D

2 Give **three** ways in which energy loss through windows and doors of a house can be reduced. **(3 marks)**

....

....

....

....

....

....

HIGHER
B

3 The pie chart below outlines the main uses of energy in the home.

(a) Using the data in the table below, complete the pie chart and the key. **(3 marks)**

> **Guided**

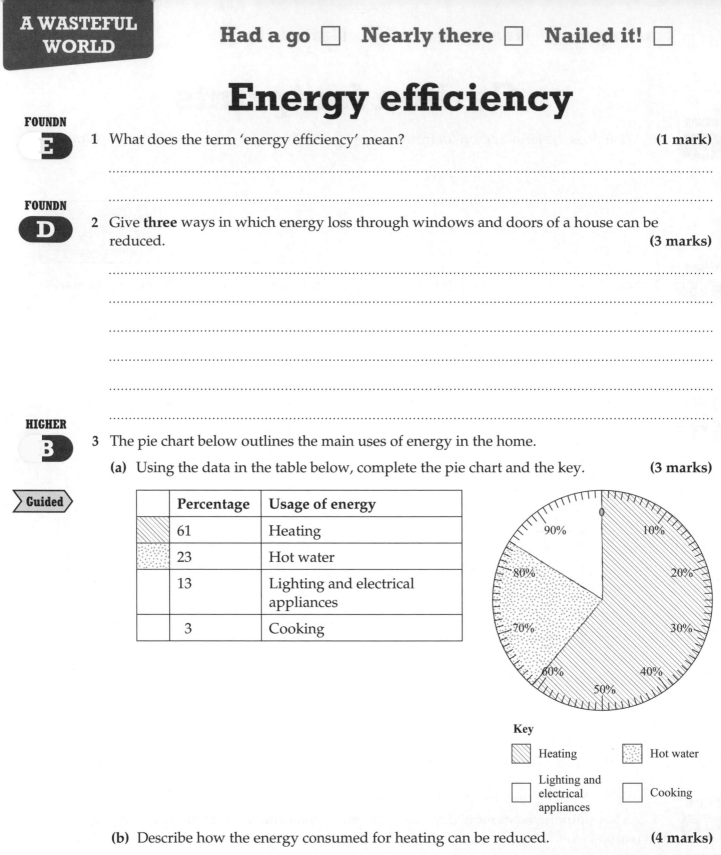

	Percentage	Usage of energy
▨	61	Heating
▦	23	Hot water
	13	Lighting and electrical appliances
	3	Cooking

Key

▨ Heating   ▦ Hot water

☐ Lighting and electrical appliances   ☐ Cooking

(b) Describe how the energy consumed for heating can be reduced. **(4 marks)**

....

....

....

....

....

....

....

....

Water consumption

FOUNDN
D

1 There are three main global uses of water. Use the data in the table to complete and label the pie chart to show these uses. **(4 marks)**

Agriculture	70%
Industry	22%
Domestic	8%

> Use a sharp pencil and ruler. Add a key!

FOUNDN
D

HIGHER
B

2 Describe **one** way water is used for each of the following: agriculture, industry and domestic reasons. **(3 marks)**

Agriculture ....

....

Industry ....

....

Domestic ....

....

HIGHER
C

3 Explain why agriculture uses 91% of total water consumed in LICs. **(2 marks)**

In LICs, farming is the biggest economic activity and therefore ....

....

....

....

> **Guided**

HIGHER
B

4 Explain why domestic supplies account for 14% of total water consumption in HICs compared to only 4% in LICs. **(4 marks)**

....

....

....

....

....

....

....

....

> Remember you are only explaining **domestic** usage, not industrial or agricultural.

Rising water use

FOUNDN D

1 How does the consumption of water change as a country becomes more developed? **(1 mark)**

....

....

FOUNDN C

2 Why is global water usage for tourism and leisure predicted to increase in the future? Include examples in your answer. **(4 marks)**

....

....

....

....

....

....

....

....

HIGHER B

3 The diagram below shows the water consumption for India in 2000 and the estimated consumption in 2050.

Billion litres per day

2000	2050 (E)
93	277
115	441
1,658	1,745

☐ Domestic ▨ Industrial ▨ Agricultural

Describe the changes in consumption shown by the diagram. Use data in your answer. **(4 marks)**

Guided

Overall water consumption is expected to increase by 597 billion/litres per day. ....

....

....

....

....

....

....

> Make sure you comment on all three water uses!

....

Local water sources

FOUNDN
D

1 The table and pictogram show sources of water for southern England. Use the information in the table to complete the pictogram.

(2 marks)

Source of water	Percentage (%)	Pictogram
Aquifers	70	💧💧💧💧💧💧
Rivers	23	💧💧
Reservoirs	7	💧

💧 = 10%

HIGHER
B

Guided

2 Explain how water is stored in an aquifer.

(4 marks)

Aquifers are layers of rock deep below the surface which store large quantities of water.

An example is ....

....

....

....

....

....

HIGHER
B

3 The graph shows information about water levels in Bewl Reservoir from April 2011.

Compare the actual and average levels between April and September 2011. Use data to help.

(4 marks)

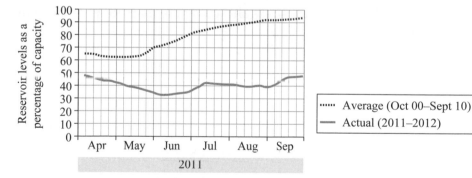

Months and years

....

....

....

....

....

....

....

Water surplus and deficit

1 The amount of water for human consumption depends on three main factors. Name the factors. **(3 marks)**

Guided

Factor 1 = r ....

Factor 2 = e ....

Factor 3 = t ....

2 What do the terms 'surplus' and 'deficit' mean? **(2 marks)**

Deficit ....

Surplus ....

3 The map shows global areas of water stress. Use the map to decide if the following statements are **true** or **false**. **(4 marks)**

True	False	
		All HICs have a water deficit
		Most of Africa has water deficit
		Australia has water surplus
		USA has a water deficit

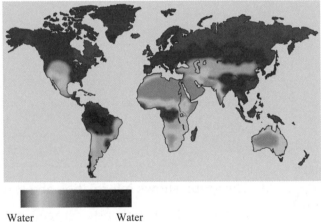

Water deficit — Water surplus

4 Describe the pattern of water stress shown by the map above. Use evidence from the map to help you. **(4 marks)**

....

....

....

....

....

....

....

5 Suggest **two** reasons why the SE of England is facing serious water stress. **(2 marks)**

....

....

....

....

Water supply problems: HICs

FOUNDN
C

1 The graph below shows the regional rainfall in the UK in August 2012 as a percentage of average August figures.

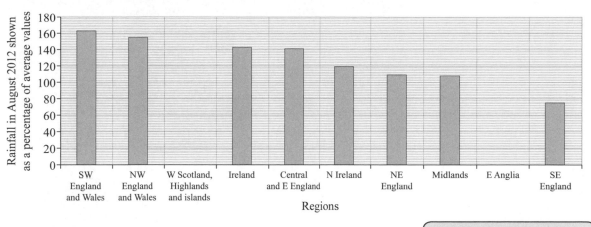

Remember any region with >100% rainfall would be **above** the normal average.

(a) Complete the graph using the data in the table below.   **(2 marks)**

W Scotland, Highlands and islands	145%
E Anglia	90%

(b) Which region is the driest?   **(1 mark)**

....

HIGHER
C

2 Suggest **two** reasons why the SE of England has water supply problems.   **(2 marks)**

....

....

....

....

HIGHER
B

Guided

3 With reference to examples, explain how water sources can be polluted.   **(4 marks)**

One of the common ways that water sources can be polluted is through industrial activity

such as mining. ....

....

....

....

....

....

....

....

Water supply problems: LICs

1 The diagram shows access to clean water for certain regions.

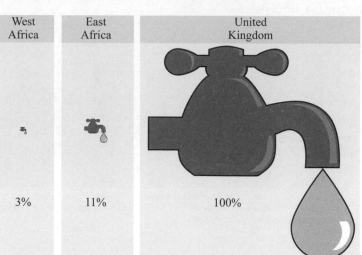

West Africa	East Africa	United Kingdom
3%	11%	100%

(a) Which region has the lowest access to clean water? **(1 mark)**

....

....

(b) How does the UK's access to clean water compare to Africa? Use data from the diagram to help. **(2 marks)**

....

....

....

....

2 WaterAid states that one child dies every 20 seconds in LICs due to waterborne diseases. Explain the reasons for high death rate. **(4 marks)**

....

....

....

....

....

....

....

....

> Exam questions similar to this have proved tricky – be prepared! ResultsPlus

Managing water in HICs

FOUNDN
D

1 Give **two** advantages to having a water meter. **(2 marks)**

Advantage 1 ....

Advantage 2 ....

FOUNDN
D

2 Approximately 8 litres of water are used to flush a toilet. Describe ways to reduce the amount of water used. **(2 marks)**

....

....

....

HIGHER
B

3 The choropleth map shows the percentage of houses in England and Wales which have a water meter fitted.

Describe the pattern shown by the map. Use data in your answer. **(4 marks)**

Less than 20%
20 – 29%
30 – 39%
40 – 49%
50% and over

....

....

....

....

....

....

....

....

....

HIGHER
B

4 Explain how the use of water for agriculture can be managed more sustainably in HICs. **(4 marks)**

Guided

Several methods can be used such as drip-feeding and irrigation sprinklers which means that

....

....

....

....

....

....

Managing water in LICs

FOUNDN
D

1 Give **two** advantages of using appropriate technology to manage water resources in LICs. **(2 marks)**

Advantage 1 ....

Advantage 2 ....

HIGHER
C

2 What is meant by the term 'rainwater harvesting'. **(2 marks)**

....

....

HIGHER
B

3 Outline the meaning of the term 'appropriate technology' when applied to managing water resources. Use an example in your answer. **(3 marks)**

....

....

....

....

....

....

FOUNDN
HIGHER
C/B

⟩ **Guided** ⟩

4 The picture shows a low-cost toilet. Give the advantages and disadvantages of using this type of appropriate technology in LICs. **(4 marks)**

Some advantages of using this type of appropriate technology in LICs include that it's very simple so can be built and maintained by local people. Also ....

....

....

....

....

....

....

Managing water with dams

FOUNDN
E

1 The divided bar graph shows how water from the Colorado River is used. Use the data in the table to complete the graph. **(3 marks)**

Percentage of water used	Usage
30%	Domestic
13%	Industrial

Percentage usage

▨	Agricultural
	Industrial
	Domestic

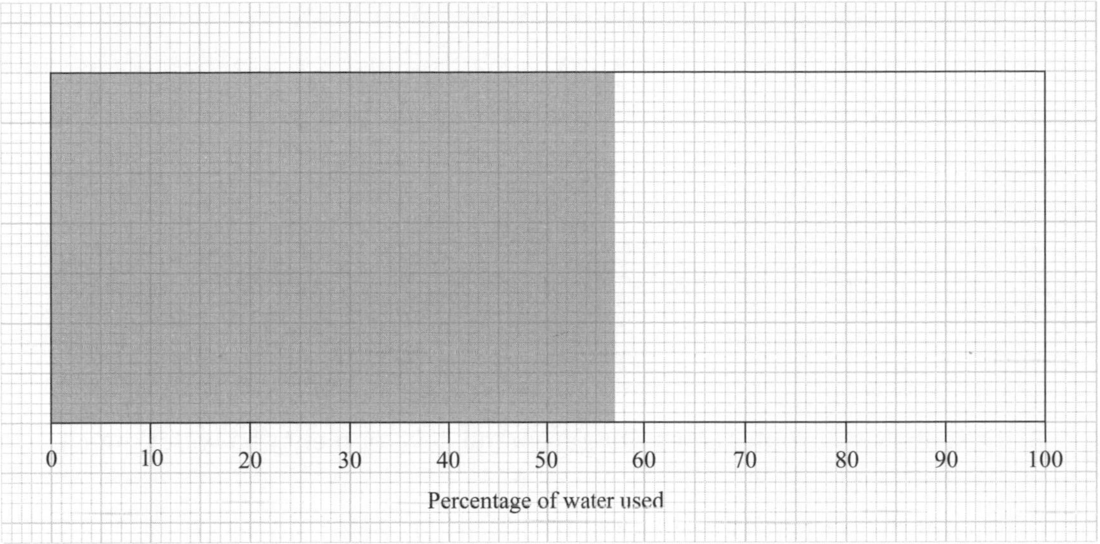

Percentage of water used

HIGHER
C

2 The increased demand for water meant that dams such as the Hoover Dam and reservoirs were built. Outline **one** advantage and **one** disadvantage of building dams. **(2 marks)**

Advantage ....

Disadvantage ....

HIGHER
A*

3 With the aid of case study detail, outline how water extraction can cause conflict.

(6 marks + 4 marks SPaG)

Guided

In the USA, dams have been built along the Colorado River which means that there is a drop in the river's flow by the time it reaches Mexico. ....

....

....

....

....

....

....

....

....

> Four marks are available for SPaG, so make sure these are really good.

Case study

WRMS, Sydney

FOUNDN
D

1 Explain what is meant by the term 'recycled water'. **(2 marks)**

....

....

....

FOUNDN
C

2 Outline **two** advantages of a water management scheme you have studied. **(4 marks)**

1 ....

....

....

....

2 ....

....

....

....

HIGHER
B

3 Describe **two** problems resulting from the development of a water management scheme you have studied. **(4 marks)**

Guided

One of the major problems with the Sydney Olympic Park scheme is that it was extremely

costly. ....

....

....

....

....

....

....

FOUNDN
C

HIGHER
A*

4 Explain how a water management scheme you have studied has helped to conserve water. **(6 marks + 4 marks SPaG)**

....

....

....

....

....

....

....

....

....

....

....

....

Economic sectors

FOUNDN
C

1 The table below shows the main economic sectors in the UK. Use the data in the table to complete the pie chart and the key. The primary sector has been plotted for you. **(3 marks)**

Economic sector	Percentage (%)
Primary	2
Secondary	18
Tertiary	78

☐ Primary

☐

☐

FOUNDN
C

Guided

2 Describe the pattern shown by the pie chart. Use data in your answer. **(2 marks)**

It is clear to see from the pie chart that the smallest sector is ....,

with ....%. An example of this industry is .... This is then

followed by ....

....

....

> Do not start the secondary data at 0%! You must start at 2%.

HIGHER
B

3 Describe how the percentage of the working population employed in each economic sector in a LIC would differ from a HIC. **(2 marks)**

....

....

....

....

> You can either start with the smallest to largest, or from largest to smallest!

HIGHER
B

**EXAM
ALERT**

4 Suggest reasons for the differences in the number of employed in the economic sectors in a HIC compared to a LIC. **(4 marks)**

....

....

....

....

....

....

....

....

> Exam questions similar to this have proved tricky – be prepared! **ResultsPlus**

Primary sector decline

FOUNDN
D

1 Mining is a primary industry that has seen a decline. State **two** reasons for this decline. **(2 marks)**

Reason 1 ....

....

Reason 2 ....

....

FOUNDN
E

2 Which of the industries in the following list can be classified as primary?

Put a tick (✓) in the correct boxes.

☐ Fishing   ☐ Forestry

☐ Car manufacturing   ☐ Catering

HIGHER
B

3 Explain how increased mechanisation impacts on primary industries. Use examples in your answer. **(4 marks)**

Guided

Mechanisation means ....

This impacts on primary industries because ....

....

....

....

....

....

....

....

> Lots of primary industries can be impacted both **positively** and **negatively** through mechanisation.

Secondary sector decline

FOUNDN
E

1   Which of the following jobs belongs to the secondary sector?   **(3 marks)**

☐ **A**   Engineer

☐ **B**   Factory worker

☐ **C**   Miner

☐ **D**   Teacher

☐ **E**   Bank clerk

☐ **F**   Shipbuilding

> The number of marks may help you decide how many to tick!

FOUNDN
E

2   Complete the sentences below by filling in the missing words using some of the words from the box.   **(4 marks)**

decline	manufacture	service	cheaper	increase
	expensive	processing	construction	

There has been a .... in secondary industry in the UK. Secondary industries .... raw materials into new products by .... them. A good example of a secondary industry is the .... industry.

HIGHER
B

3   Outline what is meant by the term 'globalisation'.   **(2 marks)**

....

....

....

....

HIGHER
A

4   18% of the UK workforce is employed in secondary industries. With the aid of examples, explain why this figure is quite low.   **(6 marks)**

Guided

The secondary sector is quite small in the UK because it's now possible for products to be made abroad because of ....

....

....

....

....

....

....

....

....

....

....

Case study

China (MIC)

FOUNDN
C

1 The graph shows the energy use in China between 1990 and 2000.

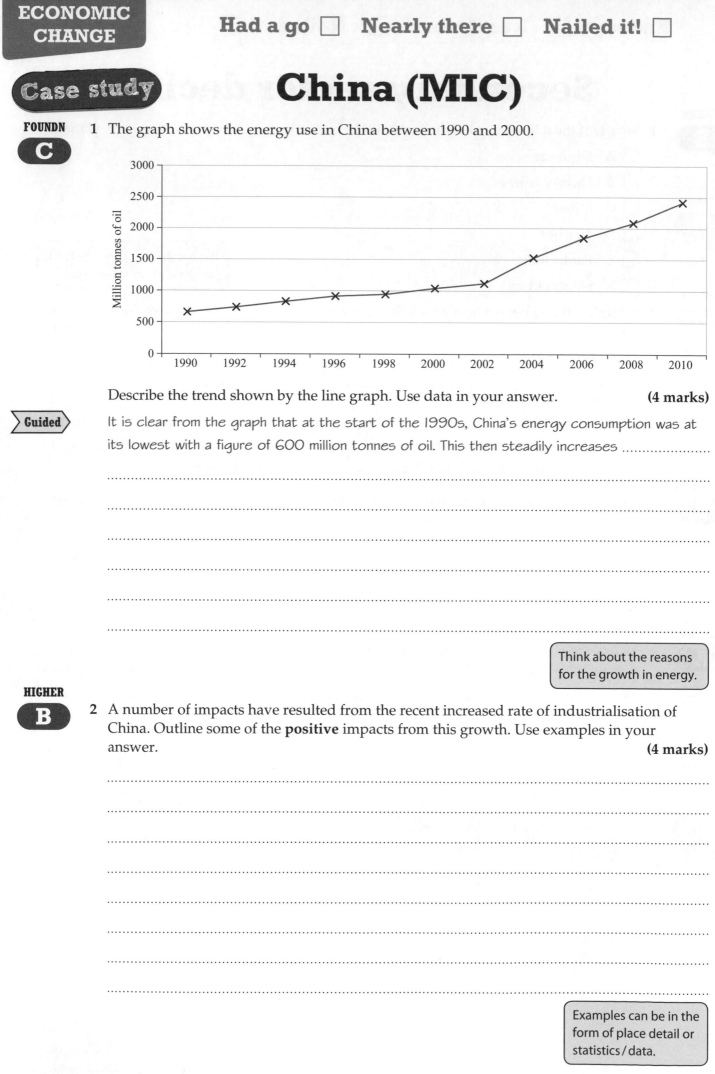

Describe the trend shown by the line graph. Use data in your answer. **(4 marks)**

> **Guided**

It is clear from the graph that at the start of the 1990s, China's energy consumption was at its lowest with a figure of 600 million tonnes of oil. This then steadily increases ....

....

....

....

....

....

....

> Think about the reasons for the growth in energy.

HIGHER
B

2 A number of impacts have resulted from the recent increased rate of industrialisation of China. Outline some of the **positive** impacts from this growth. Use examples in your answer. **(4 marks)**

....

....

....

....

....

....

....

....

> Examples can be in the form of place detail or statistics / data.

Tertiary sector growth 1

FOUNDN E

1 What does the term 'tertiary' mean? Put a cross next to the correct definition. **(1 mark)**

☐ Processing of raw materials   ☐ Production of raw materials

☐ The development of services   ☐ Research and ICT

FOUNDN D

2 Give **three** examples of a tertiary job. **(3 marks)**

Example 1 ....

Example 2 ....

Example 3 ....

HIGHER A

3 With the aid of examples, describe what happens to the amount of tertiary industry as a country moves from being a LIC to a HIC. **(4 marks)**

Guided

Countries which are classified as LICs (such as Sudan and Ethiopia) tend to have a very high percentage of primary industry (>75%) and very little secondary and tertiary. This is because ....

MICs (...) in comparison ....

....

....

Finally in HICs (...), they tend to have ....

....

....

....

> Think about the economic sectors for a LIC, MIC and HIC.

HIGHER A

4 Explain how the growth in tertiary industries has been affected by the development of new technologies. Use examples in your answer. **(4 marks)**

....

....

....

....

....

....

....

....

Tertiary sector growth 2

FOUNDN

E

1 Give **two** examples of a tertiary industry. **(2 marks)**

> It is important that you understand the meaning of key terminology.

....

....

FOUNDN

D

2 The tertiary sector has grown in HICs. What percentage of economic output does it account for? **(1 mark)**

☐ **A** More than 45%   ☐ **C** More than 60%

☐ **B** More than 50%   ☐ **D** More than 75%

FOUNDN

HIGHER

C

3 The graph below shows the growth of the tertiary industry in the UK between 1964 and 2009.

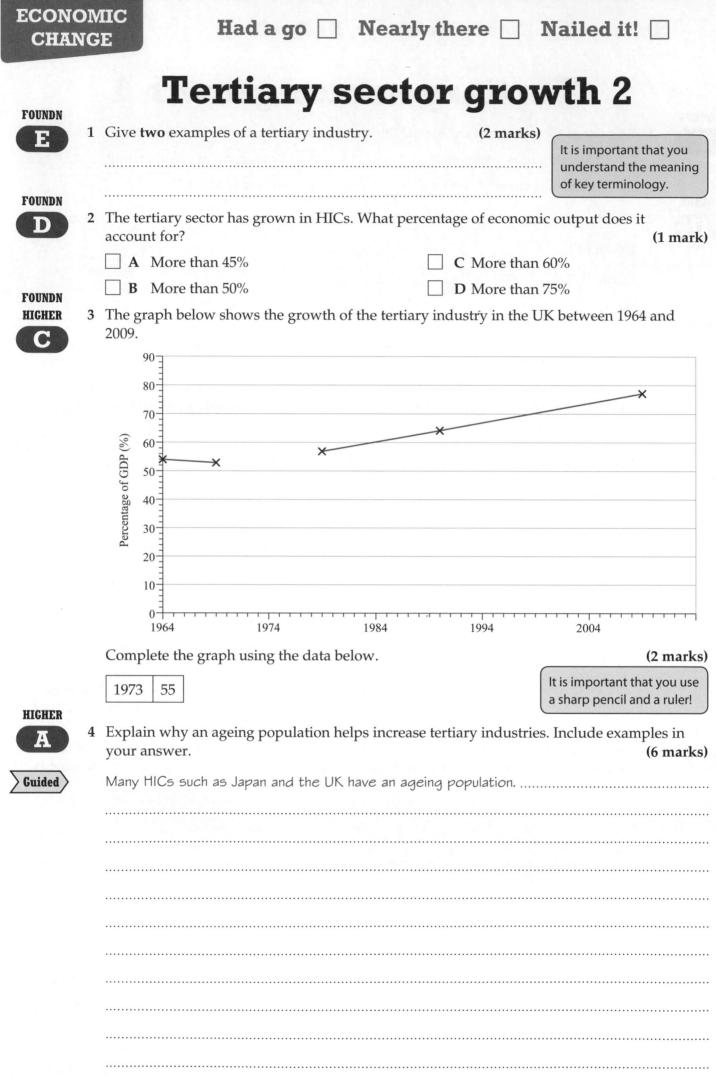

Complete the graph using the data below. **(2 marks)**

1973	55

> It is important that you use a sharp pencil and a ruler!

HIGHER

A

4 Explain why an ageing population helps increase tertiary industries. Include examples in your answer. **(6 marks)**

> Guided

Many HICs such as Japan and the UK have an ageing population. ....

....

....

....

....

....

....

....

....

Location of industries 1

FOUNDN
C

1 Give **three** socioeconomic location factors that affect the location of primary and secondary activities. **(3 marks)**

Location factor 1 ....

Location factor 2 ....

Location factor 3 ....

> Socio = people.
> Economic = money, jobs and business.

FOUNDN
D

2 Outline why iron and steel industries were frequently located close to their supplies of coal. **(2 marks)**

....

....

....

....

FOUNDN
HIGHER
C

> **Guided**

**EXAM
ALERT**

3 Explain **two physical** factors which can affect the location of primary industries. Use examples in your answer. **(4 marks)**

A primary industry such as farming is affected by physical factors. One of the main siting

factors is relief. This means the ....

This is an important location factor because ....

....

....

Another physical locating factor is ....

....

....

....

> Exam questions similar to
> this have proved tricky –
> be prepared! **ResultsPlus**

HIGHER
B

4 Outline **two** location factors affecting a secondary industry you have studied. **(4 marks)**

....

....

....

....

....

....

....

Location of industries 2

FOUNDN
D

1 What does the term 'tertiary industry' mean? **(1 mark)**

....

....

> Remember to learn
> your definitions!

FOUNDN
D

2 State **three** examples of employment which would be classified as part of the tertiary sector. **(3 marks)**

Example 1 ....

Example 2 ....

Example 3 ....

HIGHER
C

3 Why are many tertiary industries located on the edge of cities? **(1 mark)**

....

....

HIGHER
B

4 Being close to a market (source of customers) is an important siting factor for many industries.

Explain why this location factor is not important for tertiary industries. **(3 marks)**

> **Guided**

Tertiary industries do not need to be located close to their customers because of

improvements in ICT ....

....

....

....

....

HIGHER
B

5 Describe the factors affecting the location of tertiary industries. **(4 marks)**

....

....

....

....

....

....

....

> Think about specific examples
> first. E.g. what are the location
> factors for a shopping centre?

ECONOMIC CHANGE

Rural de-industrialisation: benefits

FOUNDN
C

1 What does the term 'de-industrialisation' mean? **(1 mark)**

....

....

FOUNDN
D

2 In rural communities, de-industrialisation can have many positive impacts. State **two** positive impacts. **(2 marks)**

Positive impact 1 ....

Positive impact 2 ....

HIGHER
B

3 Explain how de-industrialisation can bring benefits for the environment. **(4 marks)**

Guided

Rural de-industrialisation can have huge benefits for the environment. ....

....

....

....

It can also bring about a reduction in the amount of pollution such as ....

....

....

....

....

> Remember to focus on the **positives** for the **environment**.

HIGHER
B

4 Explain how de-industrialisation benefits housing development. **(4 marks)**

....

....

....

....

....

....

....

....

Rural de-industrialisation: costs

FOUNDN
D

1 In rural communities, de-industrialisation can have many negative impacts. Describe **two** negative impacts. **(2 marks)**

1 ....

....

2 ....

....

FOUNDN
D

2 What does the term 'rural depopulation' mean? **(1 mark)**

☐ **A** An increase in the population of urban areas

☐ **B** A decline in the number of people living in the countryside

☐ **C** When people move to the countryside from the city

☐ **D** Movement of people from one country to another

HIGHER
A

3 'De-industrialisation can be very expensive.' With the aid of examples, explain this statement. **(4 marks)**

Guided

De-industrialisation can be very expensive. The area will need to be decontaminated prior to redevelopment or rejuvenation. ....

....

....

....

....

....

....

> Think about the costs of **clearing** or **cleaning** up the area.

HIGHER
B

4 Explain why de-industrialisation can result in a rural ageing population. **(3 marks)**

....

....

....

....

....

....

....

....

Settlement functions

FOUNDN
E

1 Settlements can have a variety of functions. Match up the function to the definition. **(4 marks)**

1	Residential	**A**	Settlement based on protection from attack
2	Market town	**B**	Settlements that rely on visitors
3	Tourism	**C**	Settlements that have lots of housing
4	Strategic	**D**	Settlements that encourage trading

1 matches with ....   2 matches with ....

3 matches with ....   4 matches with ....

HIGHER
B

2 Outline how a settlement with an administrative function differs from a settlement with an industrial function. Use examples in your answer. **(4 marks)**

Guided

.... is an example of a settlement with an administrative function. ....

....

....

.... In comparison, a settlement with an industrial function is ....

....

....

....

....

> When the command word is 'differ', it means you have to make a **comparison**! To do this use a clear **connective** phrase e.g. 'in comparison to' or 'this is different from' . . .

HIGHER
B

3 With reference to an example, outline how the function of a settlement can change over time. **(4 marks)**

....

....

....

....

....

....

....

....

Change in rural communities

FOUNDN HIGHER

C

1 The graph shows the predicted population change (%) by age band for Scotland and North Ayrshire.

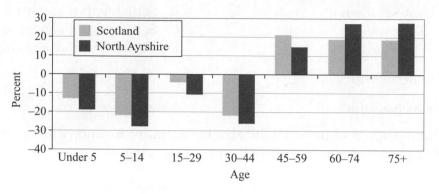

Describe the expected population decrease for the rural area of North Ayrshire. Use data in your answer. **(4 marks)**

> **Guided**

It is clear to see that there are many people leaving North Ayrshire with the biggest negative population change being in the youthful age groups 5–14 (28%). ....

....

....

....

....

....

....

> Remember to use key terms for age groups – youthful, etc.

HIGHER

B

2 Explain the reasons for the negative population growth in North Ayrshire. **(4 marks)**

....

....

....

....

....

....

....

....

HIGHER

C

3 There are many people returning to rural areas. What is the correct term for this movement? **(1 mark)**

....

....

Changing urban areas

1   The graph below shows the number of new homes built in England between 2005 and 2011.

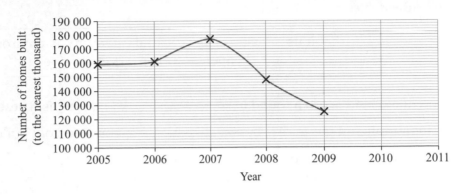

(i)   Complete the graph using the data in the table.   **(2 marks)**

2010	107 000
2011	114 000

(ii) Describe the trend shown by the graph. Use data from the graph to help.   **(4 marks)**

....

....

....

....

....

....

....

2   Outline how urban areas are changing due to de-industrialisation.   **(3 marks)**

....

....

....

....

3   Describe and explain how an increasing ageing population will impact on the construction
of new housing.   **(4 marks)**

....

....

....

....

....

....

Had a go ☐   **Nearly there** ☐   **Nailed it!** ☐

Land use change 1

1 Use some of the words in the box to complete the sentences about greenfield and brownfield sites. **(4 marks)**

habitats   electricity   rural   farm   environment
cheaper   urban   expensive

Brownfield land is often found in .... areas. These areas already have

mains supplies of gas and .... However, old buildings will need to be

removed and so developing these areas can be .... Greenfield sites are

usually found in .... areas.

2 Are the following statements **true** or **false**? Put a tick (✓) in the correct box. **(4 marks)**

True	False	
		Building on greenfield land encourages urban sprawl
		Access to services is not important when building new homes
		People are now selling off their garden as a building plot
		Eco-towns are new towns which are eco-friendly

> Exam questions similar to this have proved tricky – be prepared! **ResultsPlus**

3 Give **one** advantage and **one** disadvantage of building in the rural–urban fringe. **(2 marks)**

Advantage ....

....

Disadvantage ....

....

4 Explain the benefits of building eco-towns such as Pennbury in Leicestershire. **(6 marks)**

Eco-towns such as Pennbury help relieve the pressure on housing demands in other places,

therefore ....

....

....

....

....

....

....

....

....

....

Land use change 2

FOUNDN
D

EXAM ALERT

1 What does the term 'greenfield' mean? **(2 marks)**

....

....

....

....

> Exam questions similar to this have proved tricky – be prepared! ResultsPlus

HIGHER
B

2 Name **one** advantage and **one** disadvantage of a 'gated' community. **(2 marks)**

Advantage ....

....

Disadvantage ....

....

HIGHER
B

> Guided

3 Compare the advantages of using brownfield land as opposed to greenfield for the development of housing. **(4 marks)**

The are many advantages of using brownfield land for housing rather than greenfield land.

Brownfield sites are more likely to be located near the centre of ....

means ....

....

compared to housing in more rural greenfield sites. ....

....

....

....

....

....

> Remember that you must compare the advantages of redeveloping the **inner city** compared to greenfield land.

De-industrialisation

FOUNDN
D

1 Give **one** reason for the decline of the manufacturing industry in the UK. **(1 mark)**

....

FOUNDN
D

2 Are the following statements about de-industrialisation **true** or **false**? Put a tick (✓) in the correct box. **(4 marks)**

True	False	
		De-industrialisation often causes more jobs to be created
		There are often many brownfield sites available
		It can force people to move away from areas
		It results in many empty buildings
		Investment will be required for renewal to occur

HIGHER
A

3 Outline **two** impacts (consequences) of de-industrialisation. **(4 marks)**

....

....

....

....

....

....

....

....

> Remember that consequences can be positive **and** negative.

HIGHER
A*

Guided

4 With the aid of an example, describe and explain how the impacts of de-industrialisation can be reduced. **(6 marks)**

Areas that have suffered from de-industrialisation will have to redevelop or undergo renewal to reverse the signs of de-industrialisation. An example of an area where this has been carried out is ....

....

....

....

....

....

....

....

....

....

Brownfield and greenfield sites

1 Use some of the words in the box to complete the sentences below. **(5 marks)**

fields	cheaper	brownfield	expensive	city
urban	services	greenfield	open	housing

This is a photo of a .... site. There is lots

of .... space and ....

Developers would like to build .... in

this area as the land would be .... than

in town.

2 Use some of the words in the box to complete the sentences below. **(5 marks)**

buildings	brownfield	cheaper	expensive
expanded	derelict	cleared	empty
	greenfield	rural	

This is a photo of a .... site. There is

evidence of old .... on the site that are

.... The redevelopment of these areas is

often .... because old buildings have to be .... first.

3 Compare the costs and benefits of building on brownfield land as opposed to greenfield sites. **(4 marks)**

> Brownfield and greenfield sites both have many costs and benefits. Brownfield sites already have existing supplies of water and electricity whereas ....

....

....

....

....

....

....

....

....

> Do not be fooled by exam terminology. The costs mean the disadvantages and benefits are the advantages!

Growing cities in LICs

FOUNDN
E

1 Complete the sentences below by circling the correct answers.   **(4 marks)**

Since 1950, the population of LICs has been growing **slower / faster** than HICs. Since the 1970s, the population in the cities in LICs has been much **smaller / larger** than HICs. At present, LICs have **3× / 5× / 10×** the urban population of HICs, and by 2025 their population is expected to grow by **3 times / 5 times / 10 times**.

FOUNDN
D

2 People move into cities because of **push** and **pull** factors. State **two** push and **two** pull factors.   **(4 marks)**

Push   Pull

1 ....   1 ....

2 ....   2 ....

HIGHER
B

3 With the aid of examples, explain why in LICs, rural–urban migration occurs.   **(4 marks)**

....

....

....

....

....

....

....

....

> Remember there are rural **push** factors and urban **pull** factors to explain.

FOUNDN
HIGHER
C/B
⟩ Guided ⟩

4 Urban populations in LICs are also increasing due to a high natural increase. Explain the reasons for this.   **(4 marks)**

High natural increase occurs when the birth rates are very high and the death rates are falling.

....

....

....

....

....

....

....

Dhaka

Case study

FOUNDN
E

1 The sketch below is of a typical shanty town in a LIC.

Match the annotation in the table to A–E on the sketch. **(5 marks)**

1	Lack of access to a proper water supply	
2	Rubbish is often left at the roadside which attracts rats	
3	Some shanty towns have access to services such as electricity	
4	Very small roomed accommodation which leads to overcrowding	
5	Houses are often made out of lots of different materials	

FOUNDN
HIGHER
C

2 Why is lack of access to a proper water supply a growing problem in some urban areas of LICs? **(2 marks)**

....

....

....

....

HIGHER
B

3 With the aid of an example, outline **two** consequences caused by rapid urban growth. **(4 marks)**

Guided

One of the major problems of rapid urban growth is the strain it puts on employment.

....

....

....

....

....

....

....

....

> Think about employment, access to health and education.

Had a go ☐ **Nearly there** ☐ **Nailed it!** ☐

World population

FOUNDN
E
Guided

1 Which of the following factors can result in **dense (D)** or **sparse (S)** populations? **(5 marks)**

D	Access to a water supply
	Few natural resources
	Good trade routes
	Lots of job opportunities
	Political instability, e.g. civil war
	Climates which are not too hot or too cold

HIGHER
B

2 The graph opposite shows the global population growth between 1800 and 2027 (estimated).

It is predicted that the world population will reach 8 billion in 2027.

(a) Complete the graph by plotting the following data. **(1 mark)**

2012	7 billion

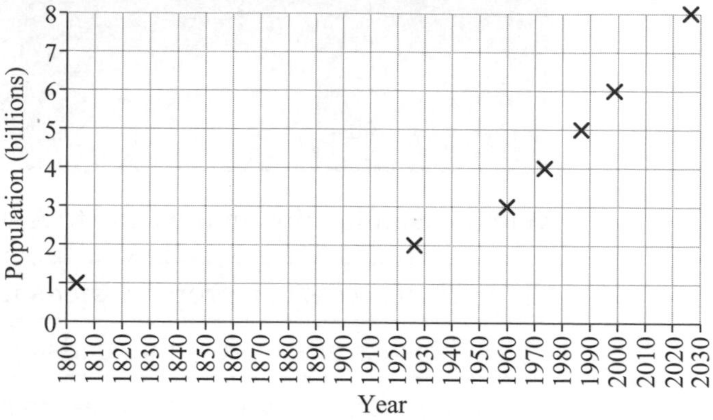

(b) Describe the trends shown by the graph. Include data in your answer. **(4 marks)**

....

....

....

....

....

....

HIGHER
A
Guided

3 Explain why some parts of the world are sparsely populated and others are densely populated. Include examples in your answer. **(6 marks)**

Population is not distributed evenly across the world. Some areas are densely populated whilst others are sparsely populated. It is densely populated in ....

....

....

....

....

....

....

....

....

....

> Remember you need to include sparsely **and** densely populated examples!

The demographic transition model

FOUNDN

D

1 Name **three** factors that affect the size of the population of a country or region.   **(3 marks)**

Factor 1 .... Factor 2 .... Factor 3 ....

FOUNDN

D

2 Using the words **high, low** or **decreasing** complete the birth rate and death rate section of the table.   **(6 marks)**

	Stage 1	Stage 2	Stage 3	Stage 4
Birth rate	High			
Death rate				Low

HIGHER

B

3 The diagram below shows a simplified version of the demographic transition model (DTM).

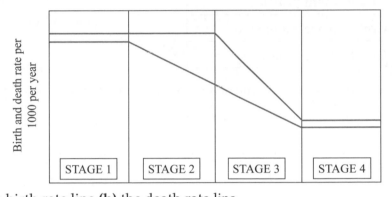

(a) Label **(a)** the birth rate line **(b)** the death rate line.   **(2 marks)**

(b) Give **three** reasons for the high birth rate shown in stages 1 and 2.   **(3 marks)**

....

....

....

....

....

HIGHER

B

4 Outline **two** reasons why birth rate drops in stage 3 of the demographic transition model. Use a named region or country in your answer.   **(4 marks)**

....

....

....

....

....

....

....

....

> Remember to include government policies as well, e.g. China and Kerala!

Birth and death rates

FOUNDN
D

1   Name another factor, other than birth and death rates, that affects the size of a country's population.   **(1 mark)**

....

FOUNDN
D

2   What is the definition of the term 'birth rate'?   **(2 marks)**

....

....

FOUNDN
E

3   Birth rates can vary over time due to **social**, **economic**, **medical** and **political** reasons.

Are the following statements linked to social (S), economic (E), medical (M) or political (P)? An example has been done for you.   **(3 marks)**

M	Little contraception
	Government policies
	Children are expensive
	Children are a status symbol

HIGHER
C

4   Outline **two** reasons why the death rate in some countries is decreasing.   **(4 marks)**

....

....

....

....

....

....

....

....

HIGHER
B

5   Explain why some named countries are experiencing a decline in the size of their populations.   **(4 marks)**

Guided

For population to decline, the birth rate has to drop below the death rate. In some countries, government policies have enforced population rules which help lower the birth rate, an example is ....

....

....

....

....

....

....

....

> Think about what happens to birth rate and death rate in a declining population.
>
> What causes this?

Population in China

1   Use some of the words in the box to complete the following sentences about population
distribution.   **(6 marks)**

| west   densely   Sahara   evenly   sparsely   Gobi   spread   east   linear   unevenly |

Population distribution is the way a population is .... out. Population is

often not distributed .... Often there are areas that have many people

(................................... populated) such as the .... coast of China, whereas

other areas have fewer people (...................................populated) such as the

.... of China.

**HIGHER
B**

2   Describe the population distribution and density of China. Use evidence from the map to
help you.   **(4 marks)**

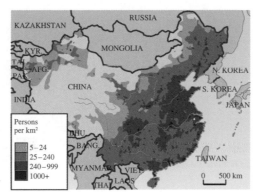

....

....

....

....

....

....

....

**FOUNDN
HIGHER
C/B**

⟩ **Guided** ⟩

3   Explain the physical factors that affect the distribution and density of China's
population.   **(4 marks)**

The .... coast is densely populated because of flat land and ....

....

....

China is sparsely populated because parts of it are very mountainous and the climate ....

....

....

....

....

Population in the UK

FOUNDN

1 Explain how the population density of the UK is influenced by physical factors. Include named examples in your answer. **(4 marks)**

> Guided

Flat land is good for farming and agriculture so flatter areas of the UK such as ....

have more people living there than mountainous areas such as ....

....

....

....

....

....

....

> Physical factors refer to mountains, rivers, soil and climate.

HIGHER

2 The choropleth map shows the population density (per km²) of the London boroughs in 2009.

Describe the pattern of distribution shown by the map.

Use evidence to support your answer. **(4 marks)**

....

....

....

....

....

....

....

....

....

....

10 000 or over
7500 – 9999
5000 – 7499
2500 – 4999
2499 or under

Enfield
Barnet
Harrow
Haringey
Waltham Forest
Redbridge
Hillingdon
Brent
1 Hackney
Camden
Newham
3
Havering
Ealing
6
7
2
4
5
11
Greenwich
Hounslow
8
9
10
12
Bexley
13
Merton
Sutton
Croydon
Bromley

1 Islington
2 Tower Hamlets
3 Barking and Dagenham
4 Hammersmith and Fulham   9 Wandsworth
5 Kensington and Chelsea   10 Lambeth
6 Westminster   11 Southwark
7 City of London   12 Lewisham
8 Richmond upon Thames   13 Kingston upon Thames

Coping with overpopulation

FOUNDN

D

1 With reference to countries you have studied, describe **three** incentives used to control birth rate. **(3 marks)**

....

....

....

....

....

....

FOUNDN

HIGHER

C

2 The graph below shows the Chinese birth rate per thousand between 1986 and 2008.

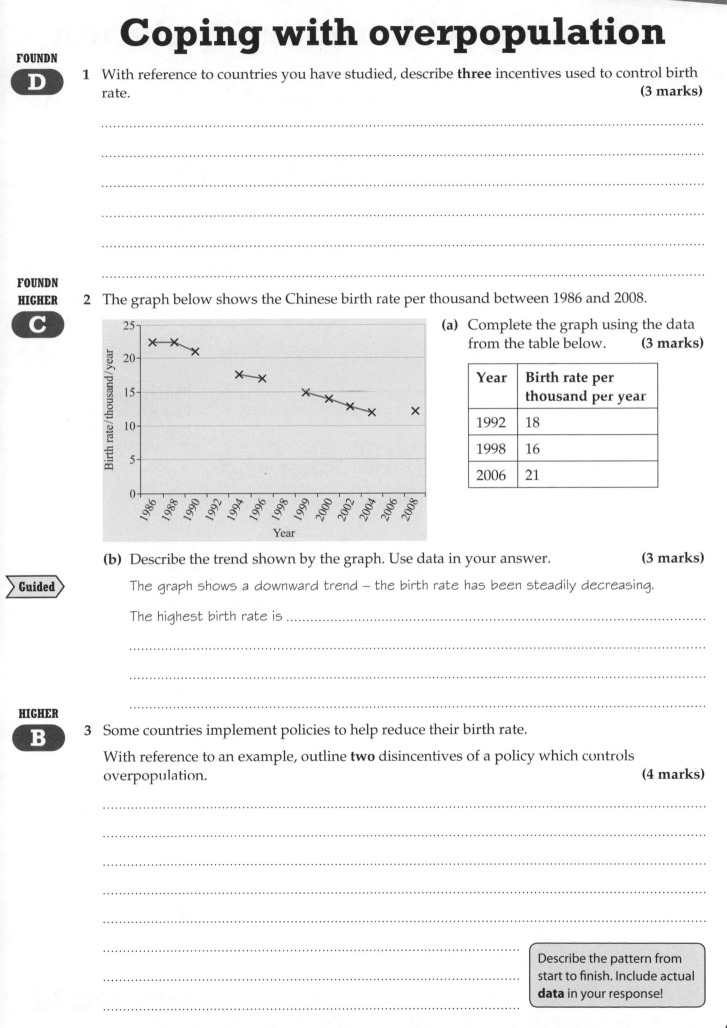

(a) Complete the graph using the data from the table below. **(3 marks)**

Year	Birth rate per thousand per year
1992	18
1998	16
2006	21

(b) Describe the trend shown by the graph. Use data in your answer. **(3 marks)**

Guided

The graph shows a downward trend – the birth rate has been steadily decreasing.

The highest birth rate is ....

....

....

....

HIGHER

B

3 Some countries implement policies to help reduce their birth rate.

With reference to an example, outline **two** disincentives of a policy which controls overpopulation. **(4 marks)**

....

....

....

....

....

....

....

> Describe the pattern from start to finish. Include actual **data** in your response!

....

Coping with underpopulation

FOUNDN

D

1   Are the following statements about over and under population **true** or **false**?

Put a tick (✓) under the correct heading for each statement.   **(5 marks)**

True	False	
		Underpopulation means that the birth rate is very low
		To help solve underpopulation emigration could be encouraged
		The 'three or more policy' was enforced to control underpopulation
		An example of a country which is underpopulated is China
		To help solve overpopulation the birth rate must increase

FOUNDN

D

2   With reference to a country you have studied, describe **one** method of increasing the birth rate.   **(2 marks)**

....

....

....

> Make sure you have the correct case study. It makes it easier to answer!

....

HIGHER

A

3   The graph shows the birth rate of women in Singapore since 1950.

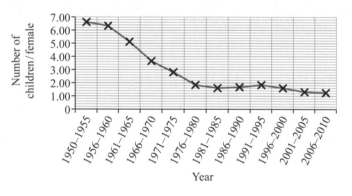

(a)   Draw an arrow onto the graph to indicate the time when Singapore become overpopulated.   **(1 mark)**

**EXAM
ALERT**

(b)   There is now concern that Singapore is underpopulated. Explain this concern using data from the graph.   **(4 marks)**

....

....

....

....

....

....

....

> Exam questions similar to this have proved tricky – be prepared! ResultsPlus

....

Population pyramids: LICs

1 Look at the graph which shows the population of a country in 2001.

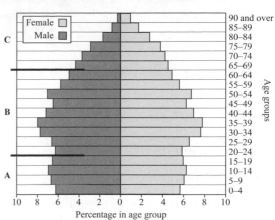

   (a) **(i)** What is this type of graph called? **(1 mark)**

....

      (ii) Which key words are best used to describe sections A, B and C on the graph?

      Circle the correct answer. **(3 marks)**

      A youthful dependents / young population

      B working population / economically active

      C ageing dependents / older population

   (b) Are the following statements about the graph **true** or **false**? **(5 marks)**

True	False	
		Infant mortality rate is low
		Life expectancy is high
		Birth rate is very high
		The graph shows a typical HIC
		This would represent the population structure of Kenya

2 Some countries have a high 'youthful dependency' ratio. What does this term mean? **(2 marks)**

....

....

....

....

3 The graph opposite shows the population pyramid of a typical LIC.

Describe, using data, the population patterns shown on the graph. **(4 marks)**

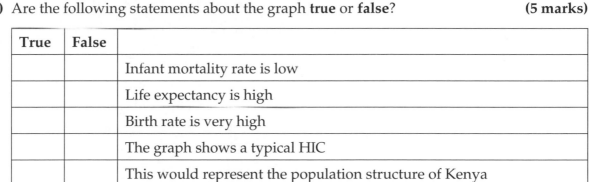

....

....

....

....

....

....

....

> Make sure you comment on birth and death rate as well as life expectancy to support your descriptions.

Population pyramids: MICs and HICs

FOUNDN
D

1 Describe **three** ways in which the population pyramid of a HIC would differ from a LIC.

(3 marks)

Difference 1 ....

Difference 2 ....

Difference 3 ....

FOUNDN
D

2 What level of development do the following population pyramids represent? Write the word LIC, MIC or HIC under the pyramids below.

(3 marks)

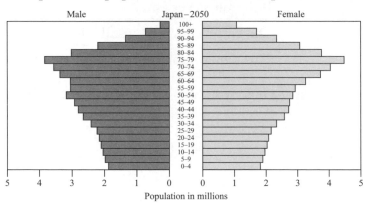

Pyramid 1   Pyramid 2   Pyramid 3

....

HIGHER
A

> Guided

3 The diagram shows the predicted population structure of Japan in 2050.

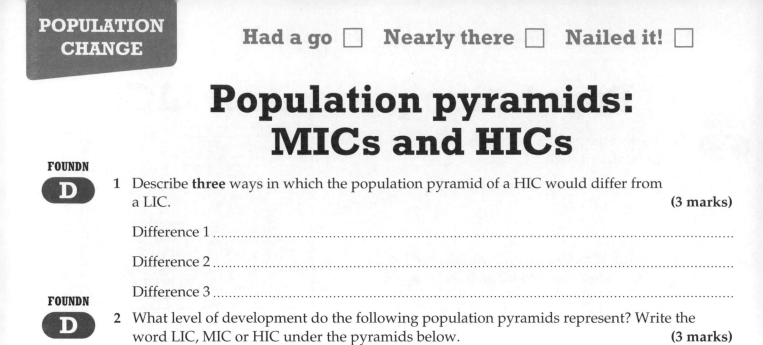

Explain the possible economic impacts of this population structure.   **(6 marks)**

One of the economic problems associated with the population pyramid is the decreased number of economically active. This is a future problem as there will not be enough workers in the population. ....

....

....

....

....

....

....

....

Young and old

FOUNDN
C

1 What does the term 'youthful population' mean? **(2 marks)**

....

....

....

FOUNDN
D

2 The birth rate in countries with a high youthful population is usually high. Suggest **two** reasons why. **(2 marks)**

1 ....

....

2 ....

....

HIGHER
B

Guided

EXAM ALERT

3 Outline **two** of the problems of a large percentage of young people (<16 years of age) in the population. Use examples in your answer. **(4 marks)**

LICs such as Malawi and Tanzania have a high youthful population. This can lead to many

problems. One such problem is ....

....

....

....

....

....

....

....

.... | Exam questions similar to this have proved tricky – be prepared! **ResultsPlus**

HIGHER
C

4 Name **two** examples of countries with an ageing population. **(2 marks)**

1 ....

2 ....

HIGHER
B

5 Suggest **two** reasons why the countries named in question 4 have high ageing populations. **(4 marks)**

....

....

....

....

....

....

....

An ageing population

FOUNDN
D

1 What does the term 'ageing population' mean? Tick (✓) the correct answer. **(1 mark)**

☐ A population with a high percentage of people under 19 years

☐ A population with a high percentage of people aged 65 or over

☐ A population with a high percentage of people aged 85 or over

☐ A population with a high percentage of people aged 65 or under

FOUNDN
D

2 State **one** advantage and **one** disadvantage of an ageing population. **(2 marks)**

Advantage ....

....

Disadvantage ....

....

HIGHER
B

3 Choose a country you have studied with an ageing population. Outline **two** costs of this ageing population. **(4 marks)**

Guided

Having a large ageing population such as .... can bring many problems.

One such problem is ....

....

....

....

....

....

....

....

> Remember costs refer to the disadvantages. Use relevant examples to support your points.

HIGHER
B

4 Explain the benefits of an ageing population. **(4 marks)**

....

....

....

....

....

....

....

....

People on the move

1 The sentences below are about population movement. Complete the sentences by filling in the blanks. Use the words in the box below to help. **(5 marks)**

| forced | migration | university student | optional | holiday in America |
| daily | flow | international | retirement to Spain | new job in south-east England |

One form of movement is .... This movement will involve a change of

home and can be voluntary or .... There are many different examples

of this type of movement, for example .... However, some people can

move for a short period of time, such as .... Other people tend to move

for a much longer period of time, e.g. ....

2 Are the following examples of migration voluntary or forced? Put a tick (✓) in the correct box. **(4 marks)**

Voluntary	Forced	
		A person leaving a country because they are a refugee
		To start a new job in another country
		Escaping from a natural hazard such as a volcanic eruption
		Leaving Manchester to live near your family in Devon

3 Explain what is meant by forced migration. Include examples in your answer. **(4 marks)**

⟩ **Guided** ⟩ Forced migration is the movement of people who have no choice but to move. One example

of this is ....

....

....

....

....

....

4 Outline the difference between net in-migration and net out-migration. **(2 marks)**

....

....

....

....

Migration into and within Europe

FOUNDN

D

1 Suggest **two** reasons why Mediterranean countries such as Italy and Greece currently have a net migration gain. **(2 marks)**

Reason 1 ....

....

Reason 2 ....

....

FOUNDN
HIGHER

C

2 The graph shows the percentage of the UK population who were born outside the UK between 1951 and 2010.

(a) Complete the graph using the data from the table. **(2 marks)**

2001	8.3%
2010 est.	11.6%

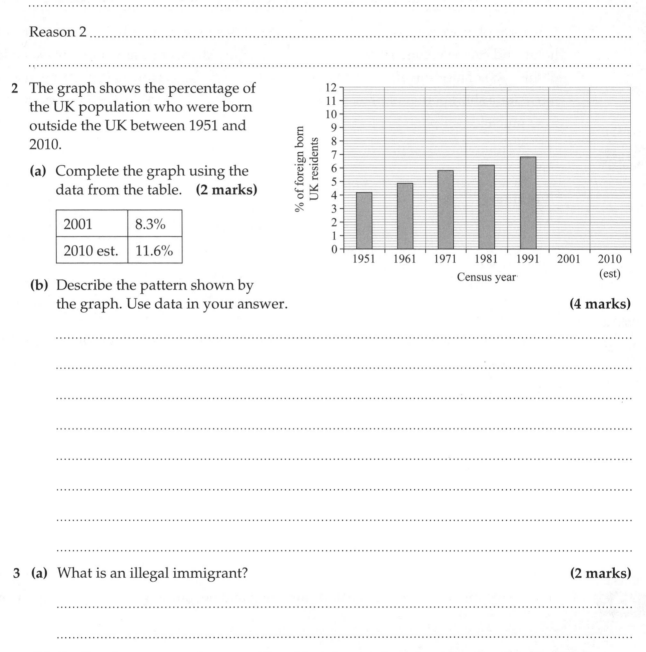

(b) Describe the pattern shown by the graph. Use data in your answer. **(4 marks)**

....

....

....

....

....

....

....

HIGHER

C

3 (a) What is an illegal immigrant? **(2 marks)**

....

....

(b) Outline the reasons why many illegal immigrants have been coming into the EU from Africa. **(3 marks)**

Guided

There are lots of illegal immigrants coming to the EU from northern parts of Africa such as Algeria and Libya. This can be voluntary ....

....

....

....

Impacts of migration

FOUNDN
C

1 Describe the social and economic consequences of migration on a named host country.

(6 marks + 4 marks SPaG)

Guided

Social consequences of migration on the host country include immigrants adding to the culture, such as through ....

....

Economic consequences of migration on the host country include ....

....

....

....

....

....

....

....

....

> An extra 4 marks are available for spelling, punctuation and grammar, so make sure these are really good. Use capital letters, full stops and specialist vocabulary accurately.

HIGHER
C

2 What is the difference between the host country and the country of origin? **(2 marks)**

....

....

....

....

HIGHER
B

3 What are the negative impacts of migration on the country of origin? Use examples in your answer. **(4 marks)**

....

....

....

....

....

....

....

Factors behind migration

FOUNDN
D

1 What is the difference between push and pull factors? **(2 marks)**

....

....

....

FOUNDN
E

2 Are the following factors **push** or **pull** factors? Put a tick (✔) in the correct box. **(4 marks)**

Push	Pull	
		Cheaper land
		Lack of medical care
		Civil war
		Higher wages

FOUNDN
D

HIGHER
B

3 Name **three** types of modern communications that help people to quickly find out information about places. **(3 marks)**

Type 1 ....

Type 2 ....

Type 3 ....

HIGHER
C

4 When national boundaries are tightened, outline what can happen to the rate of illegal migrants. **(2 marks)**

....

....

....

....

HIGHER
A*

Guided

5 Explain how modern transport has affected the rates of population movement.
Use examples in your answer. **(6 marks + 4 marks SPaG)**

With the introduction of high-speed trains such as the bullet train in Japan, people are able to travel much quicker than usual. The Channel Tunnel has increased accessibility to the Continent, in particular France. ....

....

....

....

....

....

....

....

....

Temporary migration

FOUNDN
D

1   What does the term 'short-term' population movement mean?   **(1 mark)**

....

....

FOUNDN
D

2   What is another term for short-term population movement?   **(1 mark)**

....

....

FOUNDN
HIGHER
C

⟩ **Guided** ⟩

3   In 2004, the UK received many migrants from Eastern Europe. Suggest reasons for this
    movement.   **(4 marks)**

Many migrants came to the UK from Eastern European countries such as Poland for economic

reasons. Many parts of Eastern Europe had poorer economic conditions than the UK so ....

....

....

....

....

....

....

FOUNDN
HIGHER
C/B

4   State **two** advantages of becoming a medical tourist.   **(2 marks)**

....

....

....

....

HIGHER
B

5   Outline **two** factors that influence short-term population flows for sporting reasons.
    Include an example in your answer.   **(4 marks)**

....

....

....

....

....

....

....

....

Retirement migration

FOUNDN
D

1 Complete the sentences below by filling in the blanks. Use the words in the box below to help. **(4 marks)**

sell small retirement noisy large buy quiet ageing

Many elderly people are moving. This is called .... migration. Elderly people often want a quite .... environment and their houses tend to be .... for their needs, so they downsize. Sometimes, elderly people .... their homes to help support their pension.

FOUNDN
C

2 What problems are often experienced when elderly people migrate? **(4 marks)**

....

....

....

....

....

....

....

HIGHER
B

3 Give **three** reasons why ageing populations are migrating. **(3 marks)**

....

....

....

....

....

....

FOUNDN
HIGHER
C/B
Guided
EXAM ALERT

4 Outline the consequences that retirement migration might have on the destination. **(4 marks)**

There can be positive and negative impacts of retirement migration on the destination.

The main positive impact is ....

....

However, there are many negative impacts of retirement migration. ....

....

....

....

....

....

....

Exam questions similar to this have proved tricky – be prepared! ResultsPlus

> Remember that the term 'consequence' is another way of wanting to know the **impacts**. There are often negative **and** positive consequences!

Types of 'grey' migration

FOUNDN
D

1 Name the **three** types of retirement migration.   **(3 marks)**

Retirement 1 ....

Retirement 2 ....

Retirement 3 ....

FOUNDN
C

2 Give reasons why many elderly people move into retirement apartments.   **(4 marks)**

....

....

....

....

....

....

....

HIGHER
C

3 (a) Define the term 'grey resort'.   **(2 marks)**

....

....

(b) Give one example of a 'grey resort'.   **(1 mark)**

....

....

HIGHER
B

4 The map shows the percentage of retirement migrants in regions of Spain in 2006.

Describe the pattern shown by the map.

Use evidence from the map in your answer.

(4 marks)

Values in %
☐ < 1.0
▨ 1.0 – 2.0
▩ 2.1 – 5.0
▥ 5.1 – 7.5
■ More than 7.5

— State boundary
— Boundary of autonomous region
— Provincial boundary
Galicia Name of autonomous region

....

....

....

....

....

....

....

....

Types of tourism

FOUNDN D

1 Look at the divided bar graph. It shows the purpose of visits to the UK in 2011.

Which of the following statements are true (T) or false (F)? Put a tick (✓) under the correct heading.   **(4 marks)**

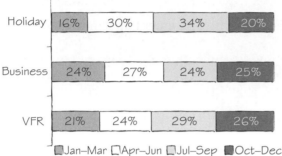

☐Jan–Mar ☐Apr–Jun ☐Jul–Sep ■Oct–Dec

	T	F
October–December is a popular time for a holiday		
General even distribution for business		
Visiting friends and relatives is least popular at the beginning of the year		
The most popular time for a holiday is the summer		

FOUNDN E

2 There are three types of tourism. Put the examples of the types of tourism in the text box under the correct type of tourism in the table.   **(3 marks)**

Conferences   Going to parents for Christmas   Meetings   Festivals
City break   Staying with a friend for the weekend

Leisure	Visiting friends and relatives (VFRs)	Business

HIGHER C

3 The divided bar graph shows the reason for tourism in the UK in 2010.

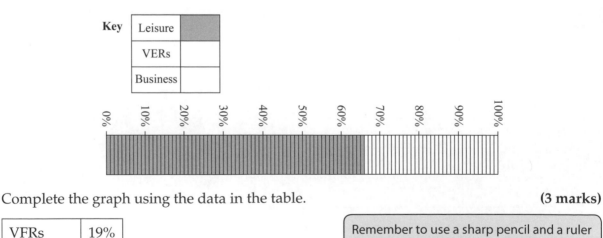

Complete the graph using the data in the table.   **(3 marks)**

VFRs	19%
Business	15%

> Remember to use a sharp pencil and a ruler for all graphical skills. Look out for whether the key needs to be completed as well!

HIGHER C

4 What does the term 'highly seasonal' mean?   **(2 marks)**

....

....

....

Why tourism is growing 1

FOUNDN
HIGHER
C

1 The line graph below shows the growth in the world's tourism since 1960.

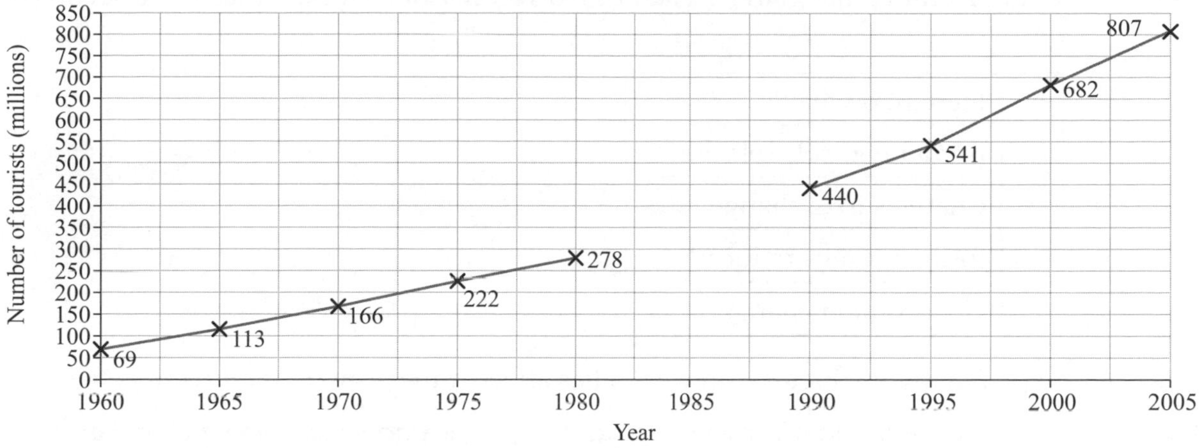

(a) Complete the graph using the data in the table.   **(2 marks)**

1985	320
1990	440

(b) How much did world tourism grow between 1960 and 1980? .... million
                                                                                        (1 mark)

(c) Which period experienced the **biggest** growth rate?   **(1 mark)**

Put a tick (✓) in the correct box

	1960–1970
	1970–1980
	1980–1990
	1990–2000

HIGHER
C

2 There are many reasons for the growth of world tourism. State **one** social reason for this growth.   **(1 mark)**

Social ....

....

HIGHER
B

3 Describe how the increase in leisure time has impacted on tourism.   **(4 marks)**

....

....

....

....

....

....

....

....

Why tourism is growing 2

FOUNDN

E

1 The following is a list of economic (E) and political (P) reasons for the growth of tourism. Using **E** and **P**, match the list and the correct reason. An example has been completed for you. **(4 marks)**

E	Increased wealth
	Border controls relaxed
	Changes in exchange rates
	Journey times reduced
	Government change

FOUNDN
HIGHER

C

⟩ **Guided** ⟩

2 Explain why the increase in wealth has affected the growth of tourism. **(4 marks)**

The increase in the minimum wage has meant that people have more disposable income to spend on luxuries such as holidays. ....

....

....

....

....

....

....

HIGHER

B

3 Explain **two** political causes for the global increase in tourism. Use examples to help support your answer. **(4 marks)**

....

....

....

....

....

....

....

....

> It is important that you RTQ – **Read the question**. It has asked for **two** causes and you must give two for the full marks.

Tourist destinations

FOUNDN
D

1 What is an adventure holiday? **(2 marks)**

....

....

FOUNDN
HIGHER
C

2 The photograph is of a holiday destination in the Austrian Alps.

Outline the physical and human attractions shown in the photograph.
(4 marks)

Physical

....

....

....

....

Human

....

....

....

....

HIGHER
B

3 The photograph was taken during the summer. Suggest how the attractions might change in this location during the winter. **(3 marks)**

Guided

During the winter, the mountains in the background of the photo would be covered in snow,

this ....

....

....

....

....

HIGHER
C

4 What is an enclave resort? **(2 marks)**

....

....

....

....

The Butler model: Blackpool

FOUNDN
C

1 Each stage of the Butler model has distinctive characteristics.

Name the stage of the model that relates to the following characteristics. **(5 marks)**

	Drop in tourist numbers
	Very few visitors
	Locals rely on tourism for income and jobs
	Area is regenerated or rebranded
	Some attractions are built

FOUNDN
E

2 Name **three** examples of attractions which could be built to increase tourist numbers to this resort. **(3 marks)**

1 ....

2 ....

3 ....

HIGHER
C

3 Explain why the introduction of trains helped to increase tourist numbers. **(4 marks)**

....

....

....

HIGHER
A*

Guided

4 Using an example of resort you have studied, explain what happened during stage 6 of the Butler model. **(6 marks + 4 marks SPaG)**

Blackpool was still highly popular as a seaside resort up until the mid 1960s. The introduction of the package holiday meant that people could visit places such as the Mediterranean more easily. ....

....

....

....

....

....

....

....

....

....

Four marks are available for SPaG so make sure your answer is well-organised.

Remember stage 6 can be divided into 3 parts: **stabilisation, decline** and **rejuvenation**.

The Butler model: Benidorm

FOUNDN
D

1 Using label lines, match up the specific statement to the stage of the Butler model. **(5 marks)**

Exploration		Many hotels and restaurants were opened
Involvement		1980s tourist numbers boomed
Development		Physical attractions were the 3 S's – sand, sea and sun
Consolidation		Other holiday destinations opened up, e.g. Turkey
Stagnation		Mayor allowed bikinis to be worn

FOUNDN
C

⟩**Guided**⟩

2 Using a resort you have studied, explain how it fits stages 1–3 of the Butler model. **(4 marks)**

In 1954, Benidorm was a town whose main function was fishing. The attractions for visitors were very much limited to the 3 S's – sea, sand and sun. ....

....

....

....

....

....

....

....

HIGHER
C

3 When tourist numbers peak, which stage of the model does this trend fit? **(1 mark)**

....

HIGHER
B

4 Describe what happens to a resort when it reaches stage 6 of the Butler model. Use an example to help. **(4 marks)**

....

....

....

....

....

....

....

> Make sure you extend your point with case study detail.

Social impacts of tourism

FOUNDN
D

1 Give **two** social impacts of tourism.   **(2 marks)**

Social 1 ....

....

Social 2 ....

....

FOUNDN
HIGHER
C

> **Guided**

EXAM ALERT

2 Describe the positive social impacts of tourism. Use examples in your answer.   **(4 marks)**

Tourism has a massive impact on people, especially those that are living close by.

....

....

....

....

....

....

....

....

....

> Exam questions similar to this have proved tricky – be prepared! **ResultsPlus**

> Make sure you stick to social impacts and do not include economic or environment impacts!

HIGHER
B

3 Outline **two** negative social impacts of mass tourism. Use examples to help you.   **(4 marks)**

....

....

....

....

....

....

....

....

Economic and environmental impacts of tourism

FOUNDN
D

1 State **one** economic and **one** environmental impact of tourism. **(2 marks)**

Economic impact ....

....

Environmental impact ....

....

FOUNDN
C
›**Guided**›

2 Describe the **positive** impacts from tourism. Use examples in your answer. **(4 marks)**

The most common positive impact from tourism is linked to the economy. In the Lake District in Cumbria, 31000 people are employed within the tourist industry.

....

....

....

....

....

....

....

....

HIGHER
B

3 Explain the negative environmental impacts of tourism. Use examples to help you. **(4 marks)**

....

....

....

....

....

....

....

....

....

Make sure you stick to the environment and do not include economic or social impacts!

Eco-tourism

FOUNDN
E

1 Use some of the words in the text box to complete the sentences below about eco-tourism. **(5 marks)**

| profits mass protect environment economy eco-friendly |
| minimise conserve jobs growth |

Eco-tourism is a form of tourism that tries to .... the damage done to the .... It also aims to provide .... for local people and ensures that the .... made stay within the area. It is a form of .... tourism.

FOUNDN
C

2 For a named eco-tourism destination explain how eco-tourism can benefit local people.
(6 marks + 4 marks SPaG)

▷ **Guided** ▷

Eco-tourism is a more environmentally friendly form of tourism and it has been successfully introduced in

> Make sure your spelling, punctuation and grammar are really good.

....

It has helped the local people by ....

....

....

....

....

....

....

....

....

HIGHER
B

3 How is eco-tourism benefiting the environment? Use examples in your answer. **(4 marks)**

....

....

....

....

....

....

> Focus on the question. Any candidates write about impacts on people with these questions!

Unit 1 Geographical Skills and Challenges

tier F

The questions listed below will make up a practice paper. This will help you practise what you have learned, but may not be representative of a real exam paper. The pages on which these questions appear are listed below.

> Time: 1 hour
> Answer **all** questions in **Section A** and **B**.

Section A

Answer **all** the questions in this section.

1. Geographical Skills

Total for question 1: 13 marks

2. Geographical Skills

Total for question 2: 6 marks

3. Geographical Skills

Total for question 3: 6 marks

Total for Section A: 25 marks

Section B

Answer **all** the questions in this section.
Spelling, punctuation and grammar (SPaG) will be assessed in question 5.

4. Challenges for the Planet

Total for question 4: 19 marks

5. Challenges for the Planet

Total for question 5: 6 marks + SPaG: 4 marks

Total for Section B: 29 marks

Total for Unit 1 Foundation paper: 54 marks

Unit 2 The Natural Environment

The questions listed below will make up a practice paper. This will help you practise what you have learned, but may not be representative of a real exam paper. The pages on which these questions appear are listed below.

Time: 1 hour 15 minutes

In **Section A** answer **all** questions in this section.

In **Section B** answer **either** question 4 **or** 5.

Section A – The Physical World

Answer **all** the questions in this section.

Total for Section A: 45 marks

Section B – Environmental Issues

Answer either question 4 **or** question 5.
Spelling, punctuation and grammar (SPaG) will be accessed in question 4 **or** 5.

Total for Section B = 24 marks

Total for Unit 2 Foundation paper: 69 marks

Unit 3 The Human Environment

The questions listed below will make up a practice paper. This will help you practise what you have learned, but may not be representative of a real exam paper. The pages on which these questions appear are listed below.

Time: 1 hour 15 minutes

In **Section A** answer **all** questions in this section.

In **Section B** answer **either** question 4 **or** 5.

Section A – The Human World

Answer **all** the questions in this section.

1. Economic Change
1	Page 82, question 2	*(1 mark)*
2	Page 78, question 1	*(2 marks)*
3	Page 76, question 1	*(4 marks)*
4	Page 81, question 2	*(2 marks)*
5	Page 79, question 2	*(2 marks)*
6	Page 79, question 3	*(4 marks)*

Total for question 1: 15 marks

2. Settlement Change
1	Page 85, question 1	*(2 marks)*
2	Page 88, question 1	*(1 mark)*
3	Page 90, question 2	*(4 marks)*
4	Page 87, question 1	*(2 marks)*
5	Page 91, question 2	*(2 marks)*
6	Page 90, question 4	*(4 marks)*

Total for question 2: 15 marks

3. Population Change
1	Page 99, question 1 (a) (i)	*(1 mark)*
2	Page 102, question 1	*(1 mark)*
3	Page 94, question 1	*(1 mark)*
4	Page 95, question 3	*(4 marks)*
5	Page 101, question 2	*(2 marks)*
6	Page 102, question 2	*(2 marks)*
7	Page 96, question 1	*(4 marks)*

Total for question 3: 15 marks

Total for Section A: 45 marks

Section B – People Issues

Answer **either** Question 4 **or** Question 5.
Spelling, punctuation and grammar (SPaG) will be accessed in question 4 **or** 5.

4. A Moving World
1	Page 104, question 1	*(2 marks)*
2(a)	Page 107, question 1	*(1 mark)*
2(b)	Page 107, question 2	*(1 mark)*
3	Page 104, question 1	*(2 marks)*
4	Page 106, question 1	*(2 marks)*
5	Page 107, question 4	*(2 marks)*
6	Page 108, question 4	*(4 marks)*
7	Page 105, question 1	*(6 marks + SPaG: 4 marks)*

Total for question 4: 20 marks + SPaG: 4 marks

5. A Tourist's World
1(a)	Page 111, question 1 (a)	*(2 marks)*
1(b)	Page 111, question 1 (b)	*(1 mark)*
1(c)	Page 111, question 1 (c)	*(1 mark)*
2	Page 116, question 1	*(2 marks)*
3	Page 117, question 1	*(2 marks)*
4	Page 113, question 1	*(2 marks)*
5	Page 112, question 2	*(4 marks)*
6	Page 118, question 2	*(6 marks + SPaG: 4 marks)*

Total for question 5: 20 marks + SPaG: 4 marks

Total for Section B = 24 marks

Total for Unit 3 Foundation paper: 69 marks

Unit 1 Geographical Skills and Challenges

The questions listed below will make up a practice paper. This will help you practise what you have learned, but may not be representative of a real exam paper. The pages on which these questions appear are listed below.

> Time: 1 hour
> Answer **all** questions in **Section A** and **B**.

Section A

Answer **all** the questions in this section.

1. Geographical Skills
1(a)	Page 5, question 3 (a)	*(1 mark)*
1(b)	Page 5, question 3 (c)	*(1 mark)*
1(c)	Page 5, question 3 (d)	*(1 mark)*
2	Page 7, question 2 (b)	*(1 mark)*
3	Page 7, question 3 (a)	*(2 marks)*
4	Page 7, question 3 (b)	*(4 marks)*

Total for question 1: 10 marks

2. Geographical Skills
1(a)	Page 10, question 2 (a)	*(1 mark)*
1(b)	Page 10, question 2 (b)	*(4 marks)*
2	Page 8, question 2	*(4 marks)*

Total for question 2: 9 marks

3. Geographical Skills
1	Page 11, question 1	*(2 marks)*
2	Page 11, question 3	*(4 marks)*

Total for question 3: 6 marks

Total for Section A: 25 marks

Section B

Answer **all** the questions in this section.
Spelling, punctuation and grammar (SPaG) will be assessed in question 5.

4. Challenges for the Planet
1	Page 12, question 2	*(3 marks)*
2	Page 12, question 3	*(4 marks)*
3	Page 13, question 3	*(4 marks)*

Total for question 4: 11 marks

5. Challenges for the Planet
1	Page 16, question 2	*(2 marks)*
2	Page 14, question 2	*(3 marks)*
3	Page 14, question 3	*(9 marks + SPaG: 4 marks)*

Total for question 5: 14 marks + SPaG: 4 marks

Total for Section B: 29 marks

Total for Unit 1 Higher paper: 54 marks

Unit 2 The Natural Environment

The questions listed below will make up a practice paper. This will help you practise what you have learned, but may not be representative of a real exam paper. The pages on which these questions appear are listed below.

> Time: 1 hour 15 minutes
>
> In **Section A** answer **all** questions in this section.
> In **Section B** answer **either** question 4 **or** 5.

Section A – The Physical World

Answer **all** the questions in this section.

1. Coastal Landscapes
1	Page 23, question 1	*(2 marks)*
2	Page 23, question 2	*(3 marks)*
3	Page 25, question 2	*(4 marks)*
4	Page 21, question 2	*(6 marks)*

Total for question 1: 15 marks

2. River Landscapes
1(a)	Page 29, question 2 (a)	*(2 marks)*
1(b)	Page 29, question 2 (b)	*(3 marks)*
2	Page 39, question 3 (b)	*(4 marks)*
3	Page 35, question 3	*(6 marks)*

Total for question 2: 15 marks

3. Tectonic Landscapes
1	Page 44, question 3	*(2 marks)*
2	Page 42, question 2	*(3 marks)*
3	Page 49, question 3	*(4 marks)*
4	Page 48, question 3	*(6 marks)*

Total for question 3: 15 marks

Total for Section A: 45 marks

Section B – Environmental Issues

Answer **either** Question 4 **or** Question 5.
Spelling, punctuation and grammar (SPaG) will be accessed in question 4 **or** 5.

4. A Wasteful World
1(a)	Page 55, question 3 (a)	*(2 marks)*
1(b)	Page 55, question 3 (b)	*(1 mark)*
2	Page 54, question 2	*(3 marks)*
3	Page 58, question 3	*(4 marks)*
4	Page 53, question 2	*(4 marks)*
5	Page 57, question 3	*(6 marks + SPaG: 4 marks)*

Total for question 4: 20 marks + SPaG: 4 marks

5. A Watery World
1	Page 63, question 2	*(3 marks)*
2	Page 70, question 3	*(3 marks)*
3	Page 68, question 2	*(4 marks)*
4	Page 70, question 4	*(4 marks)*
5	Page 71, question 3	*(6 marks + SPaG: 4 marks)*

Total for question 5: 20 marks + SPaG: 4 marks

Total for Section B = 24 marks

Total for Unit 2 Higher paper: 69 marks

Unit 3 The Human Environment

The questions listed below will make up a practice paper. This will help you practise what you have learned, but may not be representative of a real exam paper. The pages on which these questions appear are listed below.

> Time: 1 hour 15 minutes
>
> In **Section A** answer **all** questions in this section.
> In **Section B** answer **either** question 4 **or** 5.

Section A – The Human World

Answer **all** the questions in this section.

1. Economic Change

1	Page 73, question 3	*(2 marks)*
2	Page 80, question 4	*(3 marks)*
3	Page 79, question 3	*(4 marks)*
4	Page 78, question 4	*(6 marks)*

Total for question 1: 15 marks

2. Settlement Change

1	Page 86, question 3	*(2 marks)*
2	Page 85, question 2	*(3 marks)*
3	Page 90, question 4	*(4 marks)*
4	Page 86, question 4	*(6 marks)*

Total for question 2: 15 marks

3. Population Change

1	Page 99, question 2	*(2 marks)*
2	Page 97, question 2 (b)	*(3 marks)*
3	Page 94, question 5	*(4 marks)*
4	Page 100, question 3	*(6 marks)*

Total for question 3: 15 marks

Total for Section A: 45 marks

Section B – People Issues

Answer **either** Question 4 **or** Question 5.
Spelling, punctuation and grammar (SPaG) will be accessed in question 4 **or** 5.

4. A Moving World

1	Page 106, question 3	*(3 marks)*
2	Page 104, question 3 (b)	*(3 marks)*
3	Page 107, question 3	*(4 marks)*
4	Page 108, question 4	*(4 marks)*
5	Page 106, question 5	*(6 marks + SPaG: 4 marks)*

Total for question 4: 20 marks + SPaG: 4 marks

5. A Tourist's World

1	Page 113, question 4	*(2 marks)*
2	Page 111, question 1 (c)	*(1 mark)*
3	Page 113, question 3	*(3 marks)*
4	Page 116, question 2	*(4 marks)*
5	Page 112, question 2	*(4 marks)*
6	Page 114, question 4	*(6 marks + SPaG: 4 marks)*

Total for question 5: 20 marks + SPaG: 4 marks

Total for Section B = 24 marks

Total for Unit 3 Higher paper: 69 marks

1. Basic geographical skills

1 **(a)** Oblique aerial.

(b) Acceptable answers – compare different land uses, indicates height and density of settlements, outline of transport network is clear, photo available for use quicker than map, provides record of the area that day.

2 Distribution is the pattern or the way in which something exists in different amounts in different places.

3 **(a)**

(b) Emphasis is on explanatory detail.
River cliff – steep-sided bank of the river which formed when fast moving water erode the sides of the bank, causing the top of the bank to collapse into the river.

2. Cartographic skills

1 Any two from: contour lines, spot heights or shading.

2 At least two from: nucleated settlement, clustered or clumped together, located around many roadways.

3 **(a)** **(i)** Choropleth mapping.

(ii) Any two from: you can see the pattern across a large area, density patterns are easily recognisable by gradual shading, it can map quantitative data (numbered data) or qualitative data (data which shows opinions, etc.).

(b) You must describe **sparsely** and **densely** populated areas and refer to the map.
Points might include: The NE tends to be densely populated with >250 people/sq. mile. There are other pockets of dense population in the SE and SW. Central regions tend to be sparsely populated with <10 people/sq. mile. Medium population density often surrounds or is located close to areas of high density but most occurs over the eastern half of the map.

3. Sketch maps

1 North direction arrow or compass, key, scale, title.

2 Accept different shape of settlement and church with tower – example below.

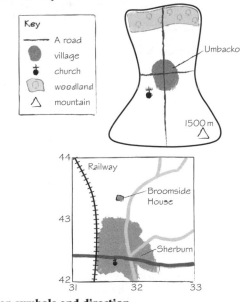

3

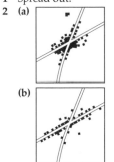

4. Map symbols and direction

1 **(a)** Parking, picnic site, viewpoint, bus station.

(b)

2 **(a)** NE.

(b) 224 m.

(c) Triangulation pillar indicates highest point of 224 m.

5. Grid references and distances

1 **(a)** Deciduous.

(b) 113911

(c) 4 km.

(d) B – Claypitts Farm.

2 C – 091923

3 **(a)** 6 km.

(b) Telephone.

(c) 112909

(d) 087919

6. Cross sections and relief

1 Patterns match as below.

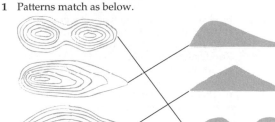

2 The land to the east of the River Otter is **steep**. It rises to a maximum height of **99 m** above sea level. In comparison the land to the west of the river is **flatter** and rises to a height of approximately **70 m** above sea level.

3 **(a)**

(b) Try to refer to the data.
The western side of the hill rises evenly in increments of 20 m. The top of the hill is relatively flat and is approximately 840 m above sea level. The downward slope off the top of the hill is a mirror image of the left (west) side. It then flattens out between 800 and 780 m and then steeply declines to 760 m.

7. Land use and settlement shapes

1 Spread out.

2 **(a)**

(b)

3 **(a)** Sidford is a nucleated settlement clustered around roads.

(b) Both physical and human land uses need to be noted and try to use map evidence.
Points to include as follows:
Physical – Sidford lies in a valley with steep land to the NE and NW; very high land (198 m) to the W; main river flows from N, tributary from NW.
Human – main settlement is Sidford; transport links into Sidford, A375 to the N; farming settlement – Brook Farm sandwiched between Core Hill and Castle Hill to the NW of Sidford.

8. Physical and human patterns

1 Must use map evidence.
- Alnwick has a nucleated settlement shape.
- It is clustered around main roads and sandwiched between the A1 and the B6341 which runs from the SW of Alnwick.

2　Must refer to map evidence or give explanation.
Alnwick will have difficulty expanding because there is coniferous woodland to the NW of the settlement, in GR 1713. This will need to be deforested to allow expansion. There is farmland to the east of the A1, therefore land is valuable for primary industry. The flow of the River Aln is to the N and NE, so will inhibit development; high relief to the W and SW which is difficult to build on.

9. Human activity on OS maps

1　Museum; campsite.
2　Any of the following: tourist information centre in GR 1813; garden in GR 1913 and castle in GR 1813.
3　Bus because there is evidence of a bus station.
4　Any two from: post office; pub; church.
5　Map evidence must be given.
Rural landscape is suggested by: a lack of built-up areas; mainly an agricultural landscape; transport routes are limited to roads less than 4 m wide which supports limited rural population; upland area and therefore landscape not able to support growth of large settlements.

10. Graphical skills

1　(a)
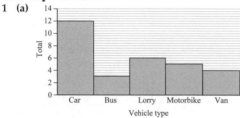

(b)　30 vehicles.
2　(a)　Population pyramid; accept age / sex diagram.
　　(b)　The country has a high birth rate – base of pyramid is very wide, lots of children in the 0–4 age band. Youthful population – pyramid narrows considerably above 30–34 age band. Top of pyramid is very narrow, indicating high death rate. Life expectancy in men is lower than in women.
3　The traffic data cannot be shown as a line graph as it does not show a change over time. Line graphs are used to show trends or patterns to see if there is a correlation between two sets of data.

11. Geographical investigation

1　Any two from: can handle large amounts of data; cover different scales; saves time; makes comparison easier; can add multiple sets of data to one map; can be amended or altered easily.
2　Aim to give at least two different methods and make sure you describe the technique.
Examples might include:
- pedestrian and traffic flow – this would help assess the impact from noise or air pollution
- environmental quality surveys – this could be used to test for litter, graffiti or vandalism
- public questionnaires – to assess public opinions on impacts from tourists or tourism.
3　You need to include evidence. Some explanation needed. Location of nearby settlements to avoid complaints about noise pollution. Height of land – high land usually has higher wind speeds. Shape of land – more rounded hills are easier for building wind farms on and easier for access to them. Also roads for access.

12. Causes of climate change

1　C – 70 ppm.
2　In 1870, carbon dioxide (CO_2) emissions were at their lowest with approximately 290 ppm. This then steadily increased until the 1950s when it reached 310 ppm. It then rapidly increased, reaching 370 ppm in 2000.
3　Try to develop your points by adding data.
HICs, such as North America and Europe are responsible for most of the CO_2 emissions due to them burning **fossil fuels** to create energy for their **industries** and also for fuel in **cars**. With increasing **car ownership**, this naturally increases CO_2.

MICs are becoming more developed in, e.g., **China** is **industrialising** and as such has increased its industry – this contributes amounts of CO_2 to the atmosphere.
Massive **deforestation** in, e.g., **Amazon, Brazil** and **Papua New Guinea**, has meant that they no longer **act as a carbon sink** and so this is therefore released into the atmosphere, increasing CO_2 levels.

13. The negative effects of climate change

1　Melting ice caps or glaciers, rising sea levels.
2　Try to give specific examples.
If temperatures increase, some areas of the world could increase in food production because the growing season will be longer and crops can be grown in higher latitudes, e.g. in Russia.
However, other areas of the world could have a decrease in food production because drier conditions will inhibit crop growth, e.g. wheat production in USA or Canada which will impact on their 43 per cent exports.
3　Try to give an explanation or detail. A range of negative impacts can be commented on – sea level change, increased incidences of coastal flooding, rising insurance claims, increased tropical storm activity. **Two** impacts need to be commented on.
One example is as follows.
The seas are heating up, which will impact on temperature-sensitive marine life such as coral. Coral reefs are being damaged by bleaching which causes them to become discoloured, turning them *white* / grey. This could then impact on the marine food chain and deplete fishing stocks and therefore could impact on fishermen.

14. Responses to climate change

1　Order – c, b, d, a.
2　Ideas could include – jacket around the boiler, not leaving electrical equipment on standby, double glazing, wall or loft insulation, turning down thermostat, etc.
3　Aim to give three or more approaches to implementation of climate change at a local scale. Make sure your points are well explained and use strong examples to illustrate your explanations. Make sure your answer is well structured and well argued and that you use appropriate geographical terminology. For the top of a level, the explanation must be balanced.
Examples of local initiatives could include – encouraging park and ride schemes, cycle paths, improvements to public transport, building of wind turbines (e.g. Samso Island, Denmark).

15. Sustainable development

1　(a)　The congestion charge is where people pay a fee to drive their vehicles into the city centres by car during peak hours.
　　(b)　Any relevant example, e.g. London, Singapore, Milan.
2　Aim to give explanation or examples.
We could introduce **park and ride**; this has been successfully achieved in **Cambridge**. This will make transport more sustainable because it will encourage people to leave their cars outside busy city centres and use public transport. This will reduce the carbon dioxide emissions and the congestion that can often occur in city centres.

16. Sustainable development: tropical rainforests

1　The answer will depend on your case study area. You need to give a range of specific points with detailed explanation and examples. This should include information on different initiatives in your chosen rainforest area. Remember that 4 marks are also available for SPaG so make sure that your spelling, punctuation and grammar are really good and that your answer is clear and well-organised. One example is as follows:
Bolivia: LICs can set aside some of their forest and receive carbon credits. Industrialised countries can then buy these credits off the country and the developing country earns money from its forest. The largest carbon credit project in the world is in Bolivia, in the Noel Kempff National Park. This is a UNESCO World Heritage Site of 1.5 million hectares and Bolivia has received £25 million by selling the carbon credits of this area. The money has gone straight to the communities who live in the area as compensation.

This means that they are no longer dependent upon logging and destroying the forest to farm to earn a living.
See answer to question 3 for other management initiatives.

2 Sustainability is about meeting the needs of the present without affecting the ability of future generations to meet their needs.

3 Try to give an example to support each of your points. Methods might include:
- preservation areas (e.g. Limonchoca Biological Reserve in the Ecuadorian Oriente)
- protection for some hardwoods, such as mahogany
- consulting local communities and establishing local land rights and ownership, to empower locals
- replanting areas to re-establish habitats
- selective logging
- educating more industrialised countries on the sustainable use of wood, e.g Japan and the TRF of Papua New Guinea
- only granting licenses to companies that buy from sustainable forestry projects
- reduce dependency on logging, oil extraction, palm oil plantations, etc. by encouraging diversification, for example to eco-tourism (Kakum National Park, Ghana) and crafts or other local products (e.g. Papua New Guinea)
- setting aside land in exchange for carbon credits (Bolivia).

17. Types of waves
1 Destructive 2 + 3.
Constructive 1 + 4.
2 Movement of water up the beach.
3 Waves hitting the SW coast of England will have **lots** of energy because the wave has travelled a **long** distance. This means that the wave will be very **powerful** and will cause lots of **erosion** at the coastline because the **backwash** is greater than the **swash**.
4 Try to give examples as well as explanation.
On the SW coast of England, the fetch is bigger because the prevailing wind travels the width of the Atlantic Ocean. This will mean that there is no land in the way to divert wave energy so waves become more powerful. In stormy conditions, wind speeds increase and this impacts on the height of surface waves, making them more powerful. This answer could also be written from the viewpoint of less powerful waves.

18. Coastal processes and landforms
1 Both weathering and erosion wear away rock. Weathering is the physical, chemical and/or biological break up of the rock in situ, whereas erosion is the movement or transport of the broken material down a slope or by water, ice or wind to another place.
2 Abrasion = Waves picking up rocks and hurling them at the cliff.
Corrosion = Dissolving of rocks and minerals in the cliff by seawater.
Hydraulic action = Water which is forced into cracks and crevices in the cliff.
Attrition = Particles become rounder as they collide within water.
3 Appropriate annotated (explanatory) comments need to be attached to the given base diagram.
The following annotations could be used.
  1 The cliff face and top are exposed to sub-aerial weathering – this includes physical, chemical and in some cases biological weathering. This makes the cliff more susceptible to collapse.
  2 The action of the sea causes erosion – mostly abrasion and attrition. This weakens the cliff base and helps with the cliff collapse.
  3 The upper part of the cliff (over the wave-cut notch) is no longer supported and will collapse into the sea.
  4 The backwash removes the collapsed cliff face and this forms a wave-cut platform.
  5 The process repeats and the cliff retreats, which increases the size of the wave-cut platform.

19. Erosional coastal landforms
1 A – Bay, B – Headland, C – Softer, less resistant rock, D – Harder, more resistant rock.
2 Make sure you use a process or full sequence.
Arches form as a result of a weakness in the cliff. Waves cause erosion through attrition and hydraulic action and cracks widen to eventually form a cave. Further erosion of the back of the cave continues until eventually the sea breaks through the back of the cave to leave a tunnel all the way through – this is called an arch.
3 Refer to both hard and soft geology.
The Isle of Wight has hard geology (chalk) which is very resistant against wave attack. However, through hydraulic action, attrition, etc. cracks in the base of the cliff can widen to form a wave-cut notch. The cliff above it becomes unstable because there is nothing to support the weight and so leads to cliff collapse.
Barton-on-Sea in Hampshire is made up of softer rock, in particular a clay layer. When it rains, the water cannot infiltrate the clay because it is impermeable. This creates a slip surface, and with wave action at the base of the cliff, slumping can occur.

20. Depositional landforms 1
1 Deposition.
2

Sediment will move along the beach	5
The angle of the wind will force longshore drift (LSD) to move in the same direction	1
The backwash will bring the beach sediment down the beach	3
A 'zig-zag' pattern occurs as the process is repeated	4
The swash will force the sediment up the beach	2

3 Try to include a process or full sequence.
Longshore drift is the movement of beach sediment along a beach. The direction of the wave depends on the prevailing wind – on the south coast, the prevailing wind is from the SW. Once the wave hits the beach, the swash will move up in the direction of the prevailing wind at a 45° angle. Once the energy has dissipated, the backwash retreats down the beach at a 90° angle, perpendicular to the beach. This process continues along the beach in a zig-zag pattern.

21. Depositional landforms 2
1 A – Bar, B – Lagoon.
2 The answer must refer to both hard and soft geology.
Softer rock, such as clays and sands, will provide less resistance than harder rock, therefore increasing the impact of the wave energy. At Swanage Bay, for example, the soft rock means that the cliffs are retreating and forming bays. Harder rock, such as Portland Stone or the Purbeck Beds, are more resistant to erosion and therefore they remain jutting out into the sea, creating headlands.

22. How coastal landforms change
1 Any of: when the cliff retreats / moves backwards / is forced inland / moves away from sea.
2 Try to include examples in your explanation.
- The longer the fetch, the more energy the wave has.
- This affects cliff recession because larger fetch results in larger waves.
- Therefore the approaching waves have more energy to erode the cliff by abrasion and hydraulic action.
- This leads to greater cliff recession.
- Whereas with a smaller fetch, the wave energy is less and so less erosion and cliff recession occur.
3 (a) Try to use the map.
The cliff has been permanently retreating since 1846. Since 1846, the cliff has retreated by about ½ km. It took 68 years, from 1887–1955, for the cliff to retreat about 250 m. From 1955–1978, the cliff retreated by about 125 m.
  (b) Softer rocks, such as muds and clays, are more susceptible to erosion; lack of coastal management techniques, such as hard engineering, to protect cliff line from wave attack; large fetch so waves more powerful therefore more erosion.

23. Coastal flooding

1 The sea wall helps prevent coastal flooding by breaking wave energy early and redirects the energy back out to sea, creating a higher barrier between the sea and the coast and stopping direct inundation of water. It also helps prevent seawater breaking onto the land during swell conditions.

2 Try to include an example.
One technique that can be used to limit damage from coastal flooding is flood shelters. These are used in poorer regions of the world, such as Bangladesh, which suffer heavily with coastal flooding because the land is very low-lying. Flood shelters protect people's livestock and possessions and so minimise the overall damage done. Other methods could be warning systems and education.

3 The answer will depend on your examples. Aim to focus on methods of forecasting, planning (to ensure preparation or defences) and building design (e.g. houses on stilts). Make sure your answer covers prevention **and** prediction. Try to explain your points really well and give a range (at least two) of specific and explained points, which could be from different examples. Make sure you include a HIC and a LIC. Examples might include the following.
- In HICs, prediction of coastal flooding is often more effective than in LICs. In the UK, the Environment Agency helps monitor weather conditions and issues relevant warnings to local councils so action can be taken. However, in LICs such as Bangladesh, prediction methods are often limited due to lack of funding and technology, therefore they tend to have prevention methods such as flood shelters and warning systems to help reduce the impact.
- HICs can afford to build flood prevention techniques such as the Thames Barrier. This will reduce the potential impacts from coastal flooding, such as the loss of property.
- Met Office in UK or work of DEFRA.

24. Coastal protection

1 Advantages – redirect wave energy back to sea; break wave energy early; often can be used as a form of promenade.
Disadvantages – very costly to build; often considered ugly; need to be maintained.

2 The answer must refer to both soft and hard techniques and include examples.
Environmentalists prefer soft engineering techniques, such as beach replenishment, because it helps prevent erosion through natural processes and often has minimal impact on the ecology of the area.
Whereas hard engineering techniques, such as groynes, are often visually polluting and can have negative impacts further down the coast.

3 (a) Groynes.
(b) You must link this to how they **prevent** erosion – don't just explain a groyne.
Groynes are used to stop the process of longshore drift. The sediment gets trapped and helps build up the beach, which acts as a natural defence against wave energy because it breaks waves early, therefore reducing their erosive power.

25. Swanage Bay and Durlston Bay

1 Bays, headlands.

2 For example, Swanage is protected by sea walls and groynes. This is because Swanage is heavily dependent on its tourism and so needs to protect the beach and tourist activities.

3 Use a located example.
In Swanage and Durlston Bay, in Dorset, they have used several techniques to protect the coastline. One way in which it is being managed is through the use of groynes. This helps protect the coastline by building up the beach at Swanage to protect the tourist industry and reduce wave attack. Another way in which it is being protected is through revetments in Durlston Bay. This helps protect the coastline by breaking wave energy early and trapping sand behind the revetment, therefore building up the beach.

4 Concentration of water facilitates drainage. This avoids moisture getting into soil/soft rock which could result in slumping.

26. The Cornish coastline

1 Prevailing wind is from the SW, therefore high fetch or there is more open water for powerful waves to build up.

2 There are two main soft engineering techniques on the Cornish coastline – dune stabilisation and beach replenishment. Any of the following two advantages: lower cost than hard engineering, low maintenance, aesthetically pleasing and can encourage wildlife and recreation.

3 The Cornish coastline is made up of very hard geology, which is therefore resistant to erosion. It has a main backbone of granite (igneous) rock, which makes it more difficult for waves to erode it and so erosional rates are much lower. The Cornish coastline erodes at a rate of 0.2–0.5 m per year.

4 Any three from: lower cost, more aesthetically pleasing therefore attractive to tourists, can encourage habitat development and low maintenance.

27. River basins

1 All rivers start in **upland** areas. They can form from bogs, lakes or **streams**. The point where a river starts is called the **source**. All rivers flow **downhill** and finally end at the **mouth** of the river.

2 A – Source, B – Mouth, C – Tributary.

3 A – Watershed, B – Mouth, C – Confluence, D – Tributary, E – Source.

28. Processes affecting river valleys

1 Abrasion – Scraping away of banks and bed by material in the water.
Corrosion – Chemicals dissolve the minerals in the rock.
Hydraulic action – Air gets squeezed into cracks in the banks and bed, forcing them apart.
Attrition – Rocks hit against each other within the river water and break up.

2 Process needs to be noted.
Water gets into cracks in the rock's surface during the day. At night, temperatures drop and the water freezes or diurnal change in temperature from warm to cold. Ice expands, which forces cracks in the rock to widen. Ice melts during daytime and the process continues until the rock particles start to fall off.

3 Answer needs to give similarities and differences (a comparison).
Both abrasion and attrition are forms of erosion but they erode material in different ways. Abrasion is the action of sandpapering on the banks and bed of the river, whereas attrition is where the rocks in the water hit against each other and become more rounded.

29. The long profile

1 (i) The width of the river **increases** downstream.
(ii) The gradient of the river decreases downstream.
(iii) The velocity of the river **increases** downstream.

2 (a) A should be V-shaped and C should be wide U-shaped.
(b) (i) B
(ii) A
(iii) C
(c) Answers need to use the shapes of the contour patterns to help justify them.
The contour lines in diagram (ii) are very close together which indicates high land where all rivers start.
Diagram (i) shows more tributaries are joining, still upland area but flatter than in A as contours are further apart.
Diagram (iii) shows much flatter land as the contours are further apart. There is also evidence of meanders.

30. Upper river landforms

1 Interlocking spur.

2 Gorge, plunge pool, granite or hard rock.

3 You need to mention the process. The full sequence is needed. A fault in the geology exposes layers of hard and soft rock. The hard rock, e.g. granite, overlies the soft rock; this is called the cap rock. Water pours over the drop, causing erosion of the softer underlying rock such as clay or sandstone. This leads to the development of a plunge pool. Overhang of harder rock eventually collapses into plunge pool. Over time, the waterfall retreats towards the source forming a steep-sided valley called a gorge.

31. Middle river landforms 1

1 A meander is a **bend** in a river. They form where the **gradient** of the river is less steep. This means that **erosion** is greatest on the outside bend where the river is **deeper**. On the inside bend, the river is slow flowing and so **deposition** occurs.

2 A – Slip-off slope, B – River cliff.

3 Fastest flow needs to be on outside bend. Slowest flow on the inside and deposition on the slip-off slope.

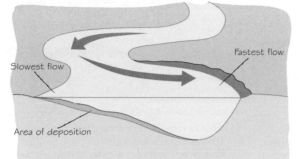

4 Lateral means **sideways** erosion or side to side.

5 Aim to include explanation as well as description.
In the middle course of the river, the gradient has become less steep. Meanders develop as riffles and pools which are alternating between shallow and deeper water in the channel. This causes the water to flow in a corkscrew formation or helicoidal flow. This results in greater erosion on the outside bend of the meander. This water movement transfers material from the outside to the inside bend and so deposition occurs on the inside bend due to a decrease in energy which aids the formation of meanders.

32. Middle river landforms 2

1 Ox-bow lake.

2 Meander, River cliff.

3 Answer should mention the process and include the full sequence. You can include an annotated diagram.
Ox-bow lakes form when rivers create meanders. Deposition occurs on the inside bend due to less erosion from hydraulic action or abrasion or corrosion. The neck of the inside bend narrows as deposition continues. This will eventually cut off the meander from the river, forming an ox-bow lake.

33. Lower river landforms

1 A – floodplain, B – meander

2 Answer should mention the process with the full sequence.
The gradient of the river is at its lowest in the lower course of the river. This slows the velocity of the river down, which means that the river doesn't have enough energy to carry its load and is therefore forced to drop or deposit the sediment. The heaviest particles are dropped first then the finer sediment later on. This can form landforms called deltas.

3 The full sequence of the process is needed. You can include an annotated diagram(s).
When a river is in spate or bankfull, water will spill onto the floodplain. The river will deposit the heaviest sediment at the sides of the river as energy is not great enough to carry it away from the river. This process continually repeats during flood events and will eventually form raised embankments called levees.

34. Why do rivers flood?

1 True, False, True, False.

2 Physical (red) – Ground was already saturated; Avon joins R. Severn; wettest June.
Human (blue) – Evacuation from Meadowhall; deforestation in Nepal.

3 The answer should include the full sequence.
The heavy storms meant that water wouldn't have been able to infiltrate into the soil quick enough. This meant there would have been excessive amounts of water on the soil surface. High land meant the water would have flowed downhill as surface run-off or overland flow and therefore enter the river system much quicker, increasing the height of the river significantly, forcing it to reach bankfull faster and then spill over.

35. River Stour, Dorset

1 Advantages – raises the height of the river; able to hold back the water from spilling onto the floodplain; can be made out of lots of different material – do not accept it prevents flooding as this is given in the question.
Disadvantages – ugly-looking; alters the flow of the river; costly to build and maintain; can burst under extreme pressure.

2 Explanation is needed with processes evident.
The deforestation of woodland would mean that there are fewer trees and leaves to intercept the rainwater, reducing interception storage. Therefore all the water hits the ground and cannot infiltrate through so is forced to flow over the ground as run-off or overland flow. Water enters the river much quicker, forcing levels of river water to rise causing flooding. The deforestation of woodland was one of the reasons for the flooding of the River Stour.

3 Your answer will depend on your chosen example. Make sure you clearly explain physical and human causes, using examples to support your points.
Examples from River Stour include the following.
• Physical – impermeable geology (e.g. clay) means that water cannot infiltrate through the soil and is forced to flow as surface run-off. This reaches the river much quicker and forces river level to rise.
• Other physical – dense network of streams, narrow floodplain, lots of rainfall.
• Human – deforestation of woodland means there are less trees and leaves to stop the water hitting the ground; this reduces interception storage. Therefore all the water hits the ground and cannot infiltrate through so it is forced to flow over the ground as run-off or overland flow. So it enters river much quicker and forces levels of river water to rise, causing flooding.

36. Mississippi, USA (2011)

1 Rivers flood for many different reasons. These can be linked to physical and / or **human** reasons. Most rivers flood because there is lots of **rainfall**, which has created excess water. Another factor is when humans build on the floodplain. This is called **land** use change. Building on the floodplain increases the number of **impermeable** (water cannot pass through) surfaces and so water run-off is much **quicker**.

2 Impacts can be linked purely to the effects on people, the environment or the economy, or a mixture of these. You should also give detail from a relevant case study. The example below uses the River Mississippi case study.
Effects of the Mississippi flood included damage to agricultural land estimated to be in the region of $2 billion, food prices increased, farmers forced to use additional fertilisers as the flooding stripped nutrients from the soil.

3 Two alternative methods linked to a case study must be explained – levees, channel stabilisation, storage areas.
For the River Mississippi, levees have been built which has increased the capacity of the river to carry more water. This then allows excess water to remain in the channel rather than flooding onto the nearby floodplains. Excessive water during flood events has been siphoned off and stored in designated areas which has reduced the water capacity in the river and reduced the chance of flooding the land.

37. Reducing the impact of flooding

1 (a) Expect flooding.
(b) Education helps minimise the damage caused. People can be informed of flooding in advance and prepare earlier. People know what to do in the event of a hazard.

2 Use map evidence.
Most chance of flooding is restricted to areas directly N and S of the river, such as Southwark. Risk of flooding decreases further inland, such as Richmond upon Thames. Areas further N and S of the river are likely to be safe, such as Islington.

3 Try to make specific points.
The Environment Agency monitors rivers and gains information on weather patterns from the Met Office in the UK. They use this information to predict areas at risk so people can be warned and take action, such as setting down sandbags to prevent their homes being flooded. Computer simulations can also be used to aid prediction, e.g. Texas medical centre in 2001 evacuated because of prediction from Rice University.

38. Flood management: soft engineering

1 Advantages – more sustainable, can often be cheaper compared to hard engineering, in keeping with natural processes, minimises impact on environment. Disadvantages – not often effective against large flood events, tends to allow the river to flood rather than stopping it.

2 Include explanation.
The replanting of trees (afforestation), especially in the upper part of the basin, increases interception storage as the leaves and branches prevent the rain from hitting the ground. The roots of the trees also take up water, this then increases filtration capacity and helps slow down the surface run-off of water, which means that it takes longer for it to reach the river and therefore reduces the chance of flooding.

3 Refer to specific groups of people – environmentalists, farmers, residents, etc.
Farmers – against: storage areas take up valuable land that could be used for farmland.
Residents – for: more aesthetically appealing than many hard engineering techniques therefore less visual pollution.
Taxpayers – for: usually less expensive to set up and maintain than hard engineering techniques.
Business developers – against: uses up land that could be developed / used for industry and business.
Environmentalists – for: more in keeping with natural processes and has minimal impact on animal habitats, etc.

39. Flood management: hard engineering

1 Using solid structures to help prevent flooding.
2 Two from – embankments, flood relief channels, straightening the river, dams.
3 (a) (i) Flood wall.
    (ii) Aim to give two reasons – increases the height of the river; river has the capacity to carry more water before flooding occurs; holds back the river water from spilling onto the floodplain.
  (b) Try to give examples which might include: Three Gorges Dam in China or Aswan High Dam in Egypt. Displacement of people, e.g. more than 1.2 million people due to the Three Gorges Dam. As a result of the creation of the reservoir lake behind the dam, habitats destroyed; reservoirs could become breeding grounds for diseases; stops the natural flooding of the river which many people rely on to refertilise the land. This could impact on farming productivity or increases incidences of parasites, e.g. bilharzia snails in Nile.

40. Earthquakes and volcanoes

1

True	False	
	✓	All volcanoes and earthquakes occur on the coast
✓		Volcanoes and earthquakes occur at plate boundaries
✓		Some volcanoes occur in the middle of plates
	✓	There are no volcanoes in South America
✓		Volcanoes and earthquakes occur in narrow bands

2 Hotspot volcano.
3 Aim to use map evidence.
The largest earthquakes (>mag 7) are located where the focal depth is the smallest, e.g. southern part of South Island; large number of earthquakes to the north of the North Island particularly to the west of the fault line; many will have quite deep focal depth (>101); few deep earthquakes in South Island; no earthquakes registered in the SE of South Island.

4 Kilauea (areas such as Yellowstone, Hawaii, Tonga, etc. would be acceptable as well).

41. Plate tectonics

1 Divergent (away), convergent (together), conservative (side by side).
2 Make sure you mention a process and include an annotated diagram.

Crust floats on the liquid magma of the mantle. Heat from the core causes convection currents to rise. As they approach the top of the mantle, the movement of the currents forces the crust to shift. The convection currents sink back towards the core. As the convection currents have become cooler and more dense on return towards the core, they heat up again. Process occurs again.

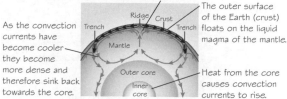

As the convection currents approach the top of the mantle the movement of the currents forces the crust to shift.

As the convection currents have become cooler they become more dense and therefore sink back towards the core.

The outer surface of the Earth (crust) floats on the liquid magma of the mantle.

Heat from the core causes convection currents to rise.

Ridge, Crust, Trench, Trench, Mantle, Outer core, Inner core

3 Answer requires an explanation of hotspots, mention of a process and the full sequence.
Hawaii is a series of volcanic islands which formed because of a hotspot in the Earth's crust. A hotspot means that the crust is thinner or weaker, which means that as magma rises from the mantle towards the surface, when it reaches the surface the magma erupts through the crust. This creates volcanoes which often rise above the ocean surface to form islands. Islands away from the source become extinct volcanoes and decrease in size due to erosion.

42. Convergent plate boundaries

1 Conservative plate margins move side by side whereas convergent plate margins move together. Earthquakes are only caused at conservative boundaries compared to both volcanoes and earthquakes at convergent boundaries.
2 Oceanic crust is more dense, younger and less thick – 6 km thick whereas continental crust can be up to 70 km thick.
3 You can include an annotated diagram.
At convergent plate margins, two plates move together and subduction of oceanic plate underneath the continental plate occurs. This can cause violent earthquakes when the plates stick and increased pressure occurs. This can then force magma to rise through volcanoes, erupting violently.

43. Divergent plate boundaries

1 False, False, True, True.
2 Order of formation:
3 The magma cools and solidifies.
1 The North American plate and Eurasian plate move apart.
5 Over time, the ridge gets bigger in size.
2 Magma rises through weaknesses in the crust.
4 This forms a ridge on the sea floor.
6 Eventually the volcano will appear above the surface of the sea.
3 At a divergent plate margin, the plates move apart. Magma rises due to convection currents, leading to pressure and doming of the crust (oceanic). Magma rises through the weaknesses (faults or fissures) in the crust. Eventually (low viscosity) magma erupts onto the surface. This cools to form volcanoes. The continued movement of plates pulls the plates apart, leading to more effusive eruptions.

44. Conservative (transform) plate boundaries

1 In a conservative plate margin the plates move **side by side**. None of the crust is being destroyed and so only **earthquakes** occur. Sometimes, the plates can get **stuck**. This can increase **pressure** which is suddenly released, causing an **earthquake**. An example of this plate margin is San Andreas Fault.
2 Transform.
3 A – North American plate.
B – Pacific plate.
4 No crust is being made or destroyed as the plates slide past each other.
5 In a conservative plate margin, plates move side by side, compared to divergent which move apart. New land or sea floor is created in a divergent but no crust is made or destroyed in a conservative. Only earthquakes occur at a conservative but volcanoes and earthquakes at a divergent.

45. Measuring earthquakes

1 Focus is the point below the ground where the earthquake starts.
2 A – focus, B – epicentre.
3 Richter and Mercalli.
4 Richter measures magnitude. The Mercalli measures damage. Richter has no upper limit. Mercalli has limit of 12. Richter is a logarithmic scale.
5 **(i)** Seismometer (not seismograph).
  (ii) Seismometers are often placed in the ground. They are sensitive to vibrations within the ground. The arm has a pen attached and once vibration is measured the pen moves and creates a seismograph, which is a graphical representation of the earthquake.

46. Living with hazards

1 Geothermal energy.
2 Two other benefits plus two examples.
  Benefits – improved soil for farming (e.g. as the slopes of Mount Etna, Sicily); minerals and precious metals such as sulphur or copper; tourism (e.g. geysers of New Zealand, Mount Vesuvius, Italy).
3 **(a)** Aim to give evidence from the map.
  For example, uneven density distribution; sparsely populated close to location of volcano (0–10 people per km²). Most densely populated on Java (over 1000 people per km²). No area on the map has maximum population density (5000+ people per km²).
  (b) Try to explain your reasons.
  The volcano has provided ash which enables fertile soil to develop. Popular tourist destination, e.g. Bali. Lack of perceived threat therefore people continue to live in the area unaware of the hazard. Lack of economic means to move. Family / emotional connection. Highly paid jobs.

47. Volcano Case study: Chances Peak volcano

1 Chances Peak is an active volcano on the island of **Montserrat** in the Caribbean. Prior to eruption, the volcano was **dormant** but became active in **1995**. The major eruption occurred on 25 **June** 1997. The main volcanic hazard was from a **pyroclastic** flow.
2 Aim to include detail and examples or data.
  Trees flattened, valleys were scraped clean of their soil, rivers became blocked with ash. This then increased chances of flooding; local wildlife killed, e.g. the eruption of Chances Peak on Montserrat had severe impacts on the environment. The pyroclastic flow resulted in 5 million m³ of rock and ash being deposited. This then blocked rivers with ash and increased the chances of flooding. The villages of Trant and Farm were covered by the ash and in total up to 4 km² of land was affected.
3 The sequence needs to be noted.
  Chances Peak in Montserrat is located on a convergent plate margin. This is where two plates are moving together, the North American and Caribbean plates. The Caribbean plate is being subducted under the North American plate (oceanic plate underneath the continental). This melts the plate, creating magma which is less dense than surrounding rock. Under increased pressure, the magma rises through the volcano and erupts.
4 Try to explain your points.
  Villages covered in ash. Two-thirds of the island covered, homes destroyed, therefore people left homeless. Some 23 people killed, burnt feet from ash. Half population evacuated to the north of the island, half population eventually left the island.

48. Earthquake Case study: Bam earthquake (2003)

1 People are unprepared. Most would still be asleep in bed, which makes it more difficult to act or evacuate.
2 Descriptive comments are acceptable for this answer. Make sure you focus on the environment.
  For Bam earthquake: fissures or cracks formed in the ground; landslides and rock falls. Ground collapsed under irrigation channels. Lack of irrigation led to loss of crops, e.g. palm and date trees dying.

3 Your answer will depend on your chosen case study. Aim to give a detailed answer, using examples from your case study to support your points. Aim to give a range of specific and explained points (these could be from different examples). Mention the short- and long-term responses.

49. Preventing tectonic hazards

1 The shape of the building is a pyramid so the base is more stable during an earthquake or reduces the centre of gravity, therefore reducing the damage caused.
2 The answer must refer to how the technique reduces damage.
  For example, automatic shutters on windows. These can prevent glass being dropped onto people in the street below. Fixing furniture to walls and floor to prevent them tipping over in earthquakes. Counterweight or any other earthquake-proofing method helps maintain the building's stability, therefore preventing it from toppling over and causing damage to transport, homes, etc.
3 One way of reducing the damage of a lava flow is to build barriers which can divert the flow of the lava. This was done at Mount Etna. This enables the flow to be safely diverted away from areas of harm, therefore reducing the overall damage caused. Explosions can also divert the flow which was also done at Mt Etna. Spraying water onto lava flow (e.g. Mt Etna) prevents lava flow from progressing towards settlements.

50. Predicting tectonic hazards

1 C – The shaking of the ground (earthquakes).
2 Aim to describe the techniques.
  Techniques: tiltmeters (changes in the angle of the slope), COSPECS (measuring gas emissions), lasers, seismometers (measure small earthquakes), satellite images (changing temperature of the surface), ultrasound (measure whether magma is rising), thermometers (measure temperature which could indicate rising magma).
  For example, scientists can measure gases given off. If there is an increase in sulphur dioxide, this could indicate magma is rising to the surface.
  Or: scientists could use satellite images which use infrared to measure surface temperature. If it is getting hotter, then magma could be rising.
3 Aim to give an explanation.
  Satellite images use infrared which measures surface temperature. If there are elevated temperatures surrounding the volcano, this could indicate that magma is rising, causing the temperature of the surface to increase.

51. A world of waste

1 **(a)**

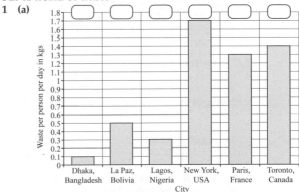

  (b) Order from left to right of the graph – L, L, L, H, H, H.
2 Two from – mobile phones, MP3 players, computers, printers, etc.
3 You need to make a comparison and use examples.
  High amounts of waste in HICs such as UK; larger amount of packaging in HICs, e.g. 8% for card and paper, glass is nearly 4%, increasing growth of e-waste / white goods. In LICs, highest waste is food, e.g. in Bangladesh, this can be up to 70%; little paper in LICs and less packaging so fewer plastics and glass.

52. Wealth and waste

1 The tendency to buy lots of consumer products and then discard them, such as throwing away a washing machine and buying another rather than repairing the old one.

2 (a) (i)

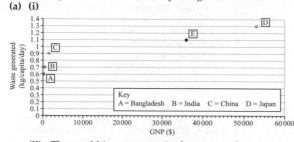

   (ii) The wealthier a country is, the greater the increase in waste produced.

   (b) Bangladesh is a very low income country with a low GDP, which means that people cannot afford to buy many products. This means less waste from packaging. Poorer regions tend to recycle and repair things such as clothes. For example, old tyres are used for a variety of purposes including sandals. There's also a low literacy rate, therefore fewer newspapers and magazines are bought.

53. Domestic waste in HICs

1 Kitchen scraps, Cardboard, Plastic, Aluminium cans.

2 Aim to use map evidence.
   Uneven distribution. There are six countries on the map which generate the highest amounts of e-waste, e.g. Sweden with 15.9 kg/person. Two countries generate the lowest amounts of e-waste per capita, e.g. Romania with 1.1 kg/person. Northern Europe tends to have the highest amounts of e-waste. Eastern Europe generally has the lowest e-waste. Western Europe generally has medium to high amounts of e-waste (4.0–8.0 kg/person)

3 Your answer will depend on your chosen examples. Aim to give a well-explained answer with a range of specific and explained points. These can be from different examples. Make sure you compare a LIC and a HIC (and explain the comparison).
   Examples could include the following.
   Higher % of household waste in UK compared to Bangladesh; higher incomes so can afford to waste products.
   Low % of paper in LIC as the literacy rate is considerably lower therefore people cannot read or afford to buy newspapers.
   Higher wastage of white goods in HICs, which fuels the consumer/throwaway society.
   HICs have a high % of food wasted. LICs practice practical recycling and so very little products are wasted.

54. Recycling waste locally

1 Reducing is the most sustainable because there is limited waste created in the first place and so less to dispose of. Does not involve changing waste products into something else.

2 Any three advantages from: recycling reduces waste going to landfill and / or incineration, therefore prevents loss of land by landfill extensions. Saves energy, saves money. Reduces the need to burn fossil fuels to create more products therefore reduces the amount of greenhouse gases emitted into the atmosphere.

3 Landfill sites are holes in the ground – either natural or man-made – where the waste is buried.

4 Advantages – large amounts of waste can be dealt with, can create electricity, can be used in heating systems.
   Disadvantages – toxins can be emitted into the air, linked with cancers, still have to landfill the ash created.

55. Recycling waste locally: Camden

1 Reduce and reuse.

2 The glass is collected and sorted at a Materials Reclamation Facility (MRF). The glass is crushed into sand and used in the construction of roads and pavements (use detail from your examples).

3 (a) 2 marks for accurate completion of the graph.

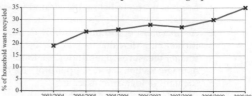

   (b) 6 per cent increase.

4 Camden in North London achieved a recycling rate of 27.2 per cent in 2007–08. To help increase this, it introduced fines. This encourages people to take part in the scheme as it was an economic incentive. Also made recycling banks available at tube stations and in their local parks to increase accessibility to recycling facilities. Operates a weekly recycling collection. Upgraded recycling centre.

56. Waste disposal in HICs

1 Greece.

2 A – Ireland, D – France, F – Portugal.

3 Your answer will depend on your chosen example. Make sure your answer is detailed and well-explained, using relevant examples to support your points. These should clearly link to the case study or example used. Remember that 4 marks are available for SPaG so make sure your answer is clear, well-organised and that your spelling, punctuation and grammar are really good. Aim to use accurate geographical terminology.
   Examples might include the following.
   Germany deals with 1/3 of its recycled waste internally and the remainder is sent abroad. About 60 per cent of all household waste is recycled and with the introduction of the green dot scheme this has meant that manufacturers have to take responsibility for the cost of recycling. This has helped reduce waste by 1 million tonnes per year and so there is less waste to deal with overall. Some regions in Germany recycle up to 67 per cent e.g. Bavaria.

57. Non-renewable energy

1 Non-renewable resources are ones that can only be used **once**. They often come from **fossil fuels**. These energy resources are often **finite**, such as **coal** and **oil**.

2 More money for the country, more foreign investment, development of infrastructure, increase in jobs for local people, help with the national debt.

3 Your answer will depend on your chosen example. Make sure your answer is detailed and well-explained, covering a range of specific points. Give relevant examples to support your points and try to give positive and negative consequences. Remember that 4 marks are available for SPaG so make sure your answer is clear, well-organised and that your spelling, punctuation, grammar and use of terminology are really good.
   Examples of consequences (Ecuador) include the following.
   Positives – Growth in economy (averaged 7 per cent annually). Makes up about 40 per cent of Ecuador's national export earnings. Increased per capita income from $290 in 1972 to $1200 in 2000. Foreign investment increased so government has become 10 times richer. The exploitation of oil was seen as a means of helping to solve Ecuador's national debt which stood at $256 million in 1970.
   Negatives – Massive deforestation. Settlers used 'slash and burn' to remove more forest and forcibly displaced indigenous tribes such as the Secoya. Very little revenue fed back into the local community. Oil pipelines dumping oil has impacted on aquatic life by reducing the amount of dissolved oxygen. Health issues linked to bathing in contaminated water. Unlined oil pits have contaminated groundwater and therefore locals have been affected. Burnt off waste caused toxic air particles which led to 'black rain' and increased greenhouse gas emissions.

58. Renewable energy

1 A – Wind, D – Solar and F – Geothermal.

2 Advantages – no greenhouse gases emitted once set up, limited impact on climate change, cheaper source of energy.
   Disadvantages – look ugly, impact on wildlife (particularly birds), often turbines built in foreign countries so limited local labour used, take up space.

3 From the map, it is clear to see that the majority of onshore turbines are located in Scotland, in areas of low population density. Very few large turbines in England, only one on the south coast of England. Many in central Wales. Largest turbines tend to be in the Highlands of Scotland, mostly in upland areas, such as Welsh mountains.

59. Energy deficit and surplus
1 More energy is produced than used.
2 False, False, True, False, True.
3 The energy it produces is equal to the energy it uses.
4 Germany has an energy deficit because it is a wealthy HIC but has limited sources of coal and oil. This means Germany spends a lot of money on products which require energy, such as cars. Germany also has a lot of industry which requires a large quantity of fuel to power them. This means they have to import large amounts of energy.

60. Wasting our energy
1 Roof – 45%, windows and doors – 35% and walls – 10%.
2 1 – Heating, 2 – Hot water, 3 – Lighting and appliances, 4 – Cooking.
3 Energy is wasted through lighting by leaving lights on when people are not in the room or not using energy-saving light bulbs.
Energy is wasted through household appliances by using inefficient or old household appliances, ill-fitting doors on fridges and freezers so cold escapes, putting dishwasher or washing on with only a half load, filling the kettle all the way up for one cup of tea.

61. Carbon footprints
1 The amount of greenhouse gases produced per capita or household or region/year, expressed as tonnes of CO_2.
2 Any three valid activities: use of public transport, cooking food, putting the heating on, etc.
3 (a)

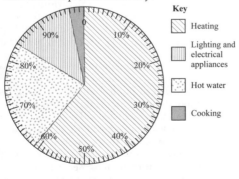

(b) **(i)** B and E.
   (ii) D and F.
4 You need to give reasons and could also give explanations. Increased development means increased disposable income. People can afford to buy more consumer products which often demand more energy, increase in industry which means more energy needed to create products, etc. Increase in waste to be disposed of or recycled, which requires energy.

62. Energy efficiency
1 Using energy in a way that minimises waste.
2 Three ways need to be given.
Installing double-glazed windows; drawing curtains or blinds over windows at night; draught-proofing doors; installing automatic door closers.
3 **(a)** Two marks for accurate completion of the pie chart. 1 mark for completion of the key.

Key
- ▨ Heating
- ▦ Lighting and electrical appliances
- ⠿ Hot water
- ▨ Cooking

(b) Reducing the thermostat by 1°C can save up to £70 per year, jacket around the boiler to stop heat escaping into the atmosphere, close the curtains at dusk – this traps heat generated by sunlight during the day inside the house, wearing thicker clothing, closing doors to keep heat inside rooms, etc.

63. Water consumption
1

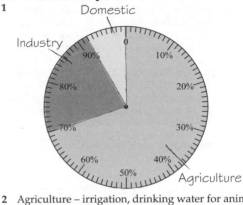

2 Agriculture – irrigation, drinking water for animals.
Industry – cooling, manufacturing processes, main ingredient for products, e.g. drinks.
Domestic – baths / showers, dishwasher, cleaning cars, watering garden, drinking.
3 Farming is the biggest economic activity in LICs therefore requires lots of water for irrigation. Inefficient use of water, leading to lots of water being wasted as run-off or left standing and so evaporates.
4 Better access to water in HICs therefore use of many labour-saving devices involving water (e.g. such as dishwashers). Sanitation is a high priority therefore people shower / wash every day, high car ownership so water used for washing. Gardens are popular. Water is used for recreation / leisure, for example swimming pools, watering golf courses, race courses, etc. LLCs have less access to domestic water supplies and less income to spend on appliances and recreation.

64. Rising water use
1 Increases.
2 Aim to give examples.
More disposable income so people have more money to spend on tourism and leisure. Can build more swimming pools, which consume large quantities of water. Golf courses need huge amounts of water for irrigation. Increase of water leisure parks, e.g. Wet n'Wild Orlando, USA.
3 Aim is to comment on the trend shown and include data in your answer.
Overall water consumption is expected to increase by 597 billion litres per day. Biggest user in each year is agriculture (1658 to 1745). Massive increase of industrial use (increase of 326 billion litres). Domestic supply is expected to increase by 2½ times.

65. Local water sources
1

Source of water	Percentage (%)	Pictogram
Aquifers	70	💧💧💧💧💧💧💧
Rivers	23	💧💧◖
Reservoirs	7	◖

2 Explanation needs to be noted.
Aquifers are layers of deep rock below the surface which store large quantities of water, e.g. Chiltern Hills or other acceptable example. When it rains, water infiltrates through the soil and is trapped by impermeable rock, e.g. granite. The stores of water are extracted through drilling wells or boreholes.
3 Aim to make a comparison to the average.
Overall usage is below the general average, lowest month is June for actual use. Average levels begin to increase from May, rising from 62 per cent to 94 per cent. From July to end of Aug, actual amount shows a minor decrease.
Highest actual amount is in Sept (48 per cent) and this is the same for the average usage (94 per cent).

66. Water surplus and deficit

1 Rainfall, evaporation and transpiration.
2 Surplus – usable water exceeds the demand. Deficit – usable water is less than demand.
3 False, True, False, False.
4 Aim to use evidence from the map.
In general, the southern hemisphere shows higher levels of water stress than the northern hemisphere. Africa and the Middle East show the highest levels of water deficit with the Amazon Basin showing the highest level of water surplus. Australia also shows high levels of water stress. The northern latitudes in general show little or no water stress..
5 Higher population densities so more demand for water, wealthier part of the country so more labour-saving devices and / or swimming pools, SE has low amounts of rainfall as it is in the rain shadow therefore less water is received to replenish reservoirs and aquifers.

67. Water supply problems: HICs

1 (a)

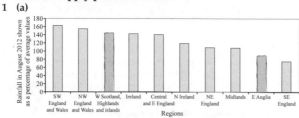

(b) SE England.
2 Reasons could include: areas where population density is generally higher therefore more demand for water. SE generally the driest region therefore area of deficit as in rain shadow of UK, therefore supply of rainfall is relatively low compared to the demand.
3 Aim to give a full explanation and include relevant examples, e.g.:
One of the common ways that water sources can be polluted is through industrial activity such as mining. In Zambia in 2006, dangerous chemicals from copper mining leaked into the local water supplies and made people ill. Poor farming practices can also cause water to be polluted. Removal of vegetation leaves soil exposed and this can be blown into rivers or use of fertilisers and pesticides can run-off into nearby water, therefore contaminating them.

68. Water supply problems: LICs

1 (a) West Africa.
  (b) All of the UK has access to clean water, compared to only 3 per cent of West and 11 per cent East Africa.
2 Limited access to drinking water so people forced to drink from contaminated supplies. Often travel great distances to reach water supplies which are often ridden with diseases such as cholera, typhoid, etc. Aim to include examples and give a full explanation.

69. Managing water in HICs

1 Reduces water bill, pay for what you use, stops excessive use of water.
2 Fitting a low flush water system, placing a hippo device into the toilet, only flush when necessary.
3 Aim to use data from the map.
Highest users of water meters are in the east and SW. There are pockets of high water users in the far SE areas of Kent where >50% of households have water metres. Lowest users are in the N of England, central London and the Midlands area with <20% of households using water metres. Coastal region of Wales has lower usage with only 20–29% having water metres.
4 You could give four brief reasons or two reasons with descriptive detail.
Drip-feed and irrigation sprinklers only used for the roots of the plants therefore water is not sitting on the soil surface so reduces evaporation rates. Rainwater harvesting storing water from storms to use during drought conditions. Water recycling using water at night to reduce the chance of evaporation.

70. Managing water in LICs

1 Provides clean water, prevents spread of disease, easily repaired, low maintenance.
2 Rainwater harvesting is the capturing of rainwater through rooftop containers.
3 Equipment that the local community is able to use relatively easily and without too much cost, e.g. boreholes or pit toilets or solar heaters.
4 Give an example to support your points. **Do not include** low cost as this is given in the question.
Advantages – strong and therefore resistant to damage from weather, permanent structure, does not have to be built with new materials, simple technology so can be built and maintained by locals.
Disadvantages – relatively expensive depending on the material used, can rust easily and may need to be replaced or maintained, can get hot during the day which could increase the smell and attract flies.

71. Managing water with dams

1

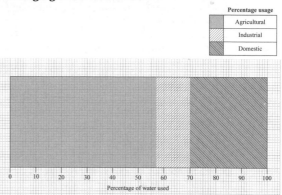

2 Advantages – reservoirs behind the dam store water for human use, recreational activities can take place on the reservoirs, HEP can be generated on some dams, controls the flow of the river therefore prevents flooding.
Disadvantages – costly to build and maintain, reservoir floods huge areas of land, habitats destroyed, reduces the flow of the river downstream.
3 Your answer will depend on your chosen case study. Aim to give a detailed and well-explained answer with a range of specific and explained points, clearly linked to your case study. Use specific examples to support your points. Four marks are available for SPaG so make sure your answer is clear, well-organised and that your spelling, punctuation, grammar and use of terminology are really good.
Examples for Colorado River include the following.
The damming of the Colorado River by the US has meant that there is a drop in the river's flow by the time it reaches Mexico. So there is less water for farmers which impacts on their productivity. The water received by Mexico is often saline which makes it unsuitable for agricultural usage, so land available for farming is reduced. The lining of the American canal has reduced water leakage but this has impacted on the aquifers in Mexico as they relied on the leakage to fill these stores of water.

72. WRMS, Sydney

1 Sewage that has been treated to remove impurities and then used for purposes other than drinking.
2 Saves drinking water for human consumption, has multiple uses, can be injected into aquifers to fill up supplies, sometimes used to prevent salt water invasion in coastal areas.
3 Cost of the scheme was very expensive, AUS$137 million to complete. Water has to be constantly monitored to ensure it is safe to use; public health is of the highest importance as it is not designed for drinking purposes.
4 Your answer will depend on your chosen example. Aim to give a detailed and well-explained answer, making a range of specific and explained points which are clearly linked to your case study or example. Use relevant examples to support your points. Remember that 4 marks are available for SPaG so make sure your answer is clear, well-organised and that your spelling, punctuation, grammar and use of terminology are really good.

Examples for Sydney Olympic Project include the following.
The Sydney Olympic Project (SOP) has helped conserve huge amounts of drinking water – 850 million litres every year. The cost of the recycled water is 15 per cent cheaper than 'normal' drinking water, which encourages the public to use it. They have recycled 100 per cent of their sewage, which has reduced the amount going to waterways and the oceans. This has significantly cut down on water pollution levels, which has helped preserve wildlife. The project has increased public awareness and education about recycled water.

73. Economic sectors

1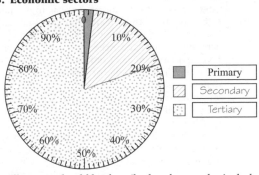

2 All sectors should be described and examples included.
It is clear to see from the pie chart that the smallest sector is primary, with only 2 per cent. An example of this industry is farming. This is then followed by secondary (such as manufacturing cars) with an increase of 9 times bigger than primary. By far the largest sector is tertiary (services such as retail) with 78 per cent.

3 Primary would be much higher with approximately 75 per cent of the working population in an average LIC. The tertiary sector is likely to be the next biggest with approximately 10 per cent of the working population. The secondary sector is likely to be limited in an average LIC.

4 Make sure you do make a comparison.
In a HIC, it is cheaper to make products abroad therefore no need to manufacture in HIC. Transport enables goods to be moved efficiently. Modern communication networks means it is easier to keep in touch. Growth of TNCs – these can operate all over the globe. De-industrialisation due to mechanisation has decreased manufacturing.

74. Primary sector decline

1 Cheaper to import than extract, cost of extraction has risen, limited supplies are left in the UK, uneconomical to run a mine, environmental implications, manufacturing industry has declined so there is a limited need for raw materials in the UK.

2 Fishing, Forestry.

3 Aim to give a good explanation of the impact of one or more forms of mechanisation.
Mechanisation means replacing human workers with machines. This impacts on primary industries because it can change the industry dramatically. For example, the fishing industry has increased mechanisation through the use of automatic net systems which can capture large numbers of fish. This unfortunately has resulted in depleted fishing stocks and in the EU placing international quotas on fish that can be caught. This then impacts on the income of fishermen. The farming industry has also mechanised through the use of tractors, bailers, etc. for arable farming and has resulted in the growth of agribusinesses or it has also resulted in factory rearing of chickens to cope with growing food demand.
Another impact of mechanisation is that it reduces the number of jobs available and so many regions have high unemployment, which forces depopulation in rural areas where most primary industries are based.

75. Secondary sector decline

1 A – Engineer, B – Factory worker, F – Shipbuilding.

2 There has been a **decline** in secondary industry in the UK. Secondary industries **manufacture** raw materials into new products by **processing** them. A good example of a secondary industry is the **construction** industry.

3 Globalisation is a process led by TNCs where the world's countries are all becoming part of one vast global economy.

4 The answer will depend on your examples. Make sure your answer is well-explained and use relevant examples to support your points. Points might include: use of modern communication systems such as email, fax, text enables people to keep in contact globally; modern transport network which can transport goods quickly and efficiently; cheaper to make product abroad therefore companies save money etc. – could also include TNCs, globalisation and de-industrialisation.

76. China (MIC)

1 Aim to include data.
It is clear from the graph that at the start of the 1990s, China's energy consumption was at its lowest with a figure of 600 million tonnes of oil. This then steadily increases until 2000 where it reaches just over 1500 million tonnes. There is a sharp increase from 2002 onwards until it reaches the highest point in 2010 with approximately 2400 million tonnes.

2 China is now the third largest economy in the world; GNP has increased, e.g. in 1995 China's GNP per capita was $816, but by 2011 this had risen to nearly $5200; raised standard of living; more money can be spent on infrastructure.

77. Tertiary sector growth 1

1 The development of services. Do not include Research and ICT as these are examples of a tertiary **industry**.

2 Any three examples of a tertiary job, e.g. teacher, nurse, electrician, builder, plumber, politician, etc.

3 Aim to describe all the movements.
Countries which are classified as LICs (such as Sudan and Ethiopia) tend to have a very high percentage of primary industry (>75 per cent) and very little secondary and tertiary. This is because there is a lack of technological advances and investment so are limited to primary. MICs (such as China) in comparison still have primary industry but their secondary and tertiary are increasing. They are becoming wealthier and so can afford to start manufacturing. China also has huge supplies of coal and raw materials to support its growing secondary industries. Finally, in HICs (such as the UK and USA), they tend to have lots of tertiary industries (>75 per cent) because they are wealthy and so can afford to import products rather than manufacture them.

4 Answer needs to include at least one reference to technology and include examples.
For example, advances in technology have meant that HICs such as the UK and USA have developed their tertiary industries using these new technologies. There are many ICT companies set up which offer a range of jobs such as internet service providers and web designing.

78. Tertiary sector growth 2

1 Two examples of tertiary industries or jobs – doctor, teacher, banker, estate agent, etc.

2 D – More than 75%.

3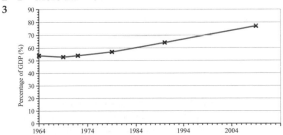

4 Your answer will depend on your examples. Make sure your answer is well-explained and give a range of specific, explained points with relevant examples to support your points. Your points can be from different examples.

79. Location of industries 1

1 Three from: near to market, local labour force, investment and capital, government policies, technology available.

2 Coal is a very heavy raw material and it is costly and difficult to transport great distances.

3 A primary industry such as farming is affected by physical factors. One of the main siting factors is relief. This means the height and slope of the land. Many farms needed extensive areas of flat land to enable arable farming, and to make it easier for tractors and bailers (mechanisation) to be used. This is why East Anglia has a lot of arable farms because it is very flat.
Another physical locating factor for farming is climate. Having an adequate rainfall and sunshine throughout the year is important to sustain crops. Arable farming cannot be done where it is too hot and dry such as the Sahara Desert.

4 Aim to make reference to an example. For example: The location factors for the Toyota car manufacturing factory in Burnaston are: large area of flat land available (100 hectares) allows the factory to expand. Large workforce available – result of high unemployment from job losses at Rolls-Royce. Energy source available from local power station. Good transport links with the A38 and M1 nearby. Local investment by council (£20 million).

80. Location of industries 2
1 The provision of a service.
2 Any three suitable jobs, e.g. teacher, dentist, doctor, shop assistant, researcher, etc. Don't use similar jobs – head teacher and teacher.
3 The land is often cheaper.
4 ICT improvements such as email, text, fax have meant customers can shop online and receive goods through deliveries; encourages e-commerce and therefore reduces the need to be located close to shops or market.
5 Explanation of siting factors is required along with relevant examples at one mark each.
As a shopping centre, the Trafford Centre has very specific siting factors. One of the main factors is that a large area of flat land is needed to enable development to take place. The land also needs to be located close to major areas of population so that you have nearby shoppers available. Often located on edge of cities, as the land is cheaper than in the city centre.

81. Rural de-industrialisation: benefits
1 Removal or decline of industry.
2 Any two from: encourages investment from different industries, sites can be used to develop much needed housing, land can be returned to farmland, ugly buildings removed, tourist attractions can develop.
3 Example **and** explanation is needed.
Rural de-industrialisation can have huge benefits for the environment. The Cotswold Water Park in Gloucestershire was once a site for the extraction of gravel. This has now been converted into a network of lakes which has increased the biodiversity of the local area, therefore benefiting wildlife, particularly birdlife. It also brings about a reduction in visual pollution as ugly buildings are removed or rejuvenated. Also, industrial discharge is reduced, e.g. slag heaps from coal mining area such as Aberfan in Wales are removed in the decontamination.
4 De-industrialisation means there is a lot of brownfield land available which can be cheaper to develop than greenfield sites. It is likely to have transport networks in place; this reduces the cost of having to build new roads and / or rail links. Services such as water, electricity and gas already in place.

82. Rural de-industrialisation: costs
1 Any two from: high unemployment, causes depopulation in the area, reduction in services, reduces the economically active workforce, etc.
2 B – A decline in the number of people living in the countryside.
3 Specific example detail is needed.
Old buildings and other structures will need to be cleared before development can begin, which can be costly. Also the area will need to be decontaminated before development which is expensive: $725 million estimate in USA. Toxic chemicals can often be left behind, e.g. sulphuric acid from gasworks in Hamburg or high concentrations of hydrocarbons from mineral oil at shipyards in Germany.

4 Usually the economically active (with families) move to urban areas as a result of unemployment in the rural areas / more employment in urban areas. This causes rural depopulation. An ageing population is often left behind as they do not need employment. May have lived there for many years and may be reluctant to change.

83. Settlement functions
1 1 – C, 2 – D, 3 – B, 4 – A.
2 Must have reference to examples and make a clear comparison.
Canberra, Australia, is an example of a settlement with an administrative function. This means the focus is on the governance of the region or country. In comparison, a settlement with an industrial function, such as Sheffield, South Yorkshire developed around a specific industrial activity, e.g. iron and steel.
3 Some explanation and example(s) must be given.
For example, Benidorm in Spain originally developed as a defensive settlement. However, its coastal location encouraged the development of a local fishing industry. It then altered its industrial activity to agricultural (particularly citrus fruits) due to the development of inland water channels. The development of main roads in the 19th century enabled visitors to come and so it became known as a tourist destination.

84. Change in rural communities
1 Aim to make reference to data from the graph.
It is clear to see that there are many people leaving North Ayrshire, with the biggest negative population change being in the youthful age groups 5–14 (28 per cent). Many of the young economically active are leaving with approximately 27 per cent moving. Overall, tends to be young families that are causing depopulation.
2 De-industrialisation of rural areas is causing a lack of jobs, particularly in the primary industries such as fishing or farming. Lack of services in rural areas such as hospitals or schools. Access to the media has educated people about urban areas. Many young people move to the cities to attend university and do not return.
3 Counter-urbanisation.

85. Changing urban areas
1 (i)

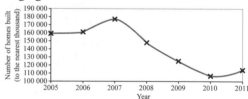

(ii) Aim to make reference to data.
From 2005 to 2007, there is an increase in the number of homes built with 18 000 more new dwellings developed. It peaks in 2007 with 177 000 then decreases sharply to 148 000 in 2008. Continues to decrease until 2010, where it reaches a low of 107 000. There is then an increase of 8 000 new homes by 2011.
2 Points might include:
- Increase in unemployment as industries have closed, this could result in counter urbanisation as people move away to find work.
- Left with derelict and possibly hazardous waste from industries, which will need to be cleaned before site can be redeveloped, this can be expensive.
- Creates brownfield sites and these could attract developers for housing projects.
3 Increased ageing population will require purpose-built flats, need mobility access close to services such as shops, banks, etc., Retirement homes needed, particular demand in popular grey areas such as the south coast.

86. Land use change 1
1 Brownfield land is often found in **urban** areas. These areas already have mains supplies of gas and **electricity**. However, old buildings will need to be removed and so developing these areas can be **expensive**. Greenfield sites are usually found in **rural** areas.
2 True, False, True, True.

3 Advantages – relatively close to the city centre, offices and factories now want to locate here – therefore job opportunities, leisure facilities and shopping complexes close by are common.
Disadvantages – built on greenfield land, so countryside lost, increases urban sprawl and estates look 'boring'.

4 Eco-towns can help relieve pressure on housing, e.g. eco-town of Pennbury helps relieve housing issues in Leicester, therefore preventing overcrowding in urban areas. Houses are often of high quality and modern, built with eco-friendly consideration such as good loft and wall insulation so limits impacts on climate change. Housing set aside for lower-income families, over 30 per cent of housing will be set aside for this. The answer will depend on your examples. Make sure your answer is well-explained and use relevant examples to support your points.

87. Land use change 2
1 Area of land on the outskirts of the city or rural areas which has never been built on before.
2 Advantages – increased security, privacy, quieter areas as surrounded on all sides.
Disadvantages – often expensive, limited to families with extensive incomes so does not address issues over low-income housing, targets for crime.
3 Some explanation is needed along with a clear comparison.
Advantages – houses are more likely to be centrally located so easier for people to get to jobs in the city centre, wide variety of shops available, leisure facilities available, reduced need to build new roads, therefore saving on destruction of habitats, reduces urban sprawl and so protects green spaces, can be cheaper for the developer as no need to include infrastructure.

88. De-industrialisation
1 Any one from: cheaper to import rather than make, labour costs cheaper in LICs, lack of local skills.
2 False, True, True, True, True.
3 Any two from: lack of jobs available; drop in services such as shops, banks, post offices; enterprising and youthful population move away; not enough economically active left to support ageing population.
4 Areas that have suffered from de-industrialisation will have to redevelop or undergo renewal to reverse the signs of de-industrialisation. One area that has done this is Bradford in West Yorkshire. Originally it was an area of textile industry but decline occurred in the second part of the 20th century. It has now developed old, derelict buildings into more modern industries such as engineering and chemical. It has also developed new tourism and leisure facilities. Old mill buildings have been regenerated into flats and museums, this has involved redeveloping the brownfield land and making it viable. The answer will depend on your examples. Make sure your answer is well-explained and use relevant examples to support your points.

89. Brownfield and greenfield sites
1 This is a photo of a **greenfield** site. There is lots of **open** space and **fields**. Developers would like to build **housing** in this area as the land would be **cheaper** than in town.
2 This is a photo of a **brownfield** site. There is evidence of old **buildings** on the site that are **derelict**. The redevelopment of these areas is often **expensive** because old buildings have to be **cleared** first.
3 Both brownfield and greenfield must be noted along with a clear comparison.
Brownfield sites already have existing supplies of water and electricity; greenfield have to put these in so costs go up. Brownfield sites are often near areas of employment, making it convenient to travel to work; greenfield can be some distance away from jobs and services. Brownfield sites reduce urban sprawl; loss of habitat in greenfield areas. Brownfield often contaminated with waste from industry which needs to be environmentally cleaned prior to development, costs a lot of money therefore increases price of development.

90. Growing cities in LICs
1 faster, larger, 3×, 5 times.
2 Push – lack of access to clean water, increased mechanisation leading to lack of jobs, lack of facilities. Pull – more manufacturing jobs, better paid jobs, better schools or medical care.
3 Both push **and** pull factors are needed plus an example(s). People leave the rural areas because of push factors such as increased mechanisation, low productivity, drought and lack of opportunity. There are limited jobs available in farming regions and people are persuaded to move to cities such as São Paulo in Brazil by pull factors such as better hospitals or schools and more and better paid jobs. This will increase the urban population.
4 Needs to refer to birth rate and death rate.
High natural increase occurs when the birth rates are very high and the death rates are falling. Birth rates in LICs are very high because their infant mortality rate is high, this leads to women having more children to compensate for those that die. Death rates in LICs are also decreasing due to improved medical care; vaccinations and jabs particularly for children help prevent common diseases.

91. Dhaka
1 1 = E, 2 = C, 3 = A, 4 = D, 5 = B.
2 Limited piped water, so people forced to drink from contaminated water supplies such as rivers, leads to disease such as cholera.
3 Any appropriate LIC example – São Paulo, Mumbai, etc. Lack of employment so people forced to work in informal sector (rubbish collection, shoe shining), lack of housing so shanty towns develop with lack of facilities, increased overcrowding as people are squeezed onto small areas of land so disease can spread quickly, increased strain on facilities such as health and education.

92. World population
1

D	Access to a water supply
S	Few natural resources
D	Good trade routes
D	Lots of job opportunities
S	Political instability, e.g. civil war
D	Climates which are not too hot or too cold

2 (a) Remember the dot needs to be slightly over 2010!

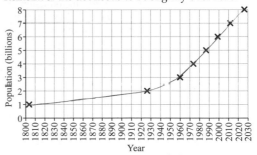

(b) Make sure you include reference to data from the graph. The lowest population was in 1800 with 1 billion. It then slowly increased until 1930 or it doubled in population in 130 years. From 1930 onwards till 1960, the population increases quickly to 3 billion. From 1960 onwards, population increases very quickly, it has doubled in 40 years. The rate of population growth increased between 2000 and 2011, where 1 billion were added in just 11 years.

3 Your answer will depend on your examples. Make sure your answer is well-explained and give a range of specific, explained points with relevant examples to support your points. Both sparsely and densely populated areas need to be described and explained. For example:
Varied physical conditions so different capacities for supporting people. Mountainous areas can lead to a sparse population, example – Rockies or Alps because of poor soil and / or challenging climate. Infertile soil can lead to a sparse population because low carrying capacity or equivalent idea. Areas near the coast (for example Cardiff) can lead to a dense population as they are a good location

for a port to develop. Industry attracts people as a workforce as in south Wales. Tourism may lead to growth of population as in Gower peninsula.

93. The demographic transition model

1 Birth rate, death rate and migration.

2

	Stage 1	Stage 2	Stage 3	Stage 4
Birth rate	High	High	Decreasing	Low
Death rate	High	Decreasing	Decreasing	Low

3 **(a)** Two marks for correctly identifying and labelling the lines.

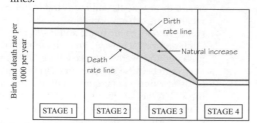

(b) Lack of contraception, high infant mortality rate so have more babies to compensate, lack of medical care for expectant mothers and babies, economic asset, to look after parents when older.

4 Could be taken from the viewpoint of any country – HIC such as Spain, Germany or from the point of view of a country trying to actively decrease birth rate – China. Later marriages, so tend to have fewer babies. Women have careers therefore do not have babies until much later. Government policy, e.g. one-child policy, e.g. China, sterilisation actively encouraged, e.g. Kerala, India.

94. Birth and death rates

1 Migration.
2 The number of **live** births per thousand / per year.
3 Government policies (P), children are expensive (E) and status symbol (S).
4 Infant mortality rate is lower. Improved health care, e.g. jabs or inoculations such as typhoid. Access to clean water supply so less incidences of disease such as cholera. Sewage is being disposed of properly.
5 Explanation is needed plus the use of examples.
Any factors which lower birth rates such as: enforced government policies which help to lower the birth rate, e.g. China's one-child policy. Introduction of contraception or sterilisation which helps to lower the birth rate. Lower infant mortality rate so women don't have lots of children to compensate for those who previously would have died. Women have careers therefore getting married later and having fewer children as a result. Ageing population so death rate may be higher than birth rate, e.g. Germany.

95. Population in China

1 Population distribution is the way a population is **spread** out. Population is often not distributed **evenly**. Often there are areas that have many people (**densely** populated) such as the **east** coast of China, whereas other areas have fewer people (**sparsely** populated) such as the **west** of China.
2 Information given should be specific (linked to map of China).
Population is uneven distributed, densely populated on the east / south coast, e.g. Shanghai or Beijing (1000+ people per km²). Sparsely populated towards the north / west (5–24 people per km²). Population decreases from east to west, from 1000+ people to 5 people per km².
3 Aim to give specific (linked to a specific area) or developed points with explanation.
East coast is densely populated because of flat land mostly under 500 m above sea level, near rivers, temperate climate. Western China is sparsely populated because it is mountainous (in the south west) and inhospitable climate, e.g. Gobi desert (in the north west). Very dry <500 mm rainfall/year. Try to avoid reversal ideas – east is flat whereas west is mountainous.

96. Population in the UK

1 Needs to be explained and have examples. Focus on physical factors.
Flat land, good for farming and agriculture such as southern England. Rainfall helps create fertile soils for farming and water supply for human use, such as the south west. Mountainous land is less fertile and difficult to farm so fewer people live in mountainous areas such as the Highlands of Scotland.
2 Population density is uneven, most densely populated areas are located around the central area of the borough, located along the river – populations of 10 000 people or more, or example such as Hammersmith and Fulham. Sparsely populated areas located at the edges – 2499 people or fewer, or example such as Bromley. Moderate density tends to be located around the high-density areas: 5000–7499, or example such as Brent and Ealing.

97. Coping with overpopulation

1 Credit explanations if given but can still get maximum marks with descriptive points as long as **three** techniques are described.
Forced by government policy plus example such as China's one child policy, monetary incentive plus example such as Singapore, longer maternity or paternity leave, three months' leave for mothers, three days' leave for fathers for the first four children, better childcare.
2 **(a)** Three marks for correct location of the points. Line needs to join up the points – otherwise two marks.

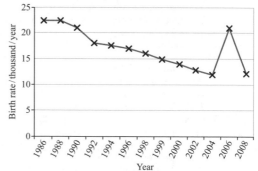

(b) Overall downward trend or relatively steady decrease, highest birth rate is in 1986 of between 22 and 23 per thousand dropped by 10 per thousand in 22 years. Sudden increase in birth rate in 2006 to 21 per thousand but sharp drop to 13 per thousand in 2008.
3 China's one-child policy: delayed marriage, abortion, sterilisation or contraception, pledge not to have children, sacked from job, fined, monitoring by the granny police.

98. Coping with underpopulation

1 True, False, True, False, False.
2 In Singapore, methods of increasing the birth rate included incentives – CDA (Child Development Account), encouraging immigration of young couples, monetary incentives, cheaper nurseries, spacious apartments, preferential schools, etc. which all tried to encourage people to have children and more of them.
3 **(a)**

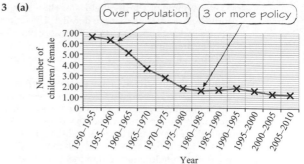

(b) The fertility rate has been dropping since 1950; it dropped from 6.6 children / female to 1.8 children / female in 1985. This drop took 35 years. Since 1985, the number of children / female has remained at around 1.4 to 1.8.

99. Population pyramids: LICs

1 (a) (i) Population pyramid or age–sex diagram.
      (ii) A – youthful dependents, B – economically active,
           C – ageing dependents.
   (b) False, True, False, True, False.
2 A high number of young dependents below the age of 16 years compared to the economically active.
3 Must refer to the graph. Comments could include: high birth rate or wide base, approximately 18.2 per cent of the population 0–4 years of age. High death rate or narrower top, approximately 0.5 per cent 80–84 years of age. Decreasing number of economically active indicates low life expectancy.

100. Population pyramids: MICs and HICs

1 Differences can be from a HIC or LIC point of view.
  HIC – wider top therefore death rate is lower and life expectancy higher; smaller base therefore birth rate is lower; wider middle therefore more economically active people.
2 Pyramid 1 = MIC, 2 = HIC, 3 = LIC.
3 Impacts can be positive and / or negative.
  Shows an ageing population – more people over 65 due to a low birth rate for the last 50 years. This mean that there are not enough economically active to support the ageing population, results in lack of money for pensions. More money needed for medical care of the elderly, e.g. nursing homes. As there is a decreased number of economically active, taxes may have to increase to cope with the high amount of ageing dependents. Ageing population spends 'greying yen' – increased business for retail and leisure, e.g. Saga holidays which cater for elderly.

101. Young and old

1 High percentage of young people in the population <16 years, accept <19 years.
2 High youthful population means high birth rate, lack of contraception, high infant mortality rate so have babies to compensate, children seen as an economic asset so want to have more, have more children to look after parents when older.
3 Must use at least one example.
  Problems: high dependency ratio, lack of economically active people to support young population, pressure on schools and paediatric medical care. Food and water supplies under pressure, growing future pressure on provision of jobs.
4 Any two suitable HICs, e.g. Japan, Germany, UK, US, Singapore.
5 More disposable income, can afford to buy higher quality food for better standard of living and nutrition. Improved medical care, health and safety laws preventing accidents. Improved sanitation, increased access to clean drinking water therefore less likely to die from diseases such as cholera, typhoid, etc.

102. An ageing population

1 A population with a high percentage of people aged 65 or over.
2 One mark for an advantage and 1 mark for a disadvantage.
  Advantage – SKI (Spending Kids' Inheritance) money welcomed by retailers, take part in more leisure activities, look after grandchildren so children can work.
  Disadvantage – pressure on pension system, lack of workforce, increased money needed for medical care, special housing needs to be built to cope.
3 Needs an example – UK, Germany, Japan, USA, etc.
  Pressure on pension system, lack of workforce, increased money needed for medical care, special housing needs to be built to cope, increase in care workers, deprivation could occur.
4 Many have paid off mortgages so available disposable income means they can spend money on leisure facilities, this then benefits UK retailers. Growth in overseas 'SKI'. Lots of LIC destinations now viewed as attractive. Retirement resorts develop in popular locations which can then benefit local retailers.

103. People on the move

1 One form of movement is **migration**. This movement will involve a change of home and can be voluntary or **forced**. There are many different examples of this type of movement, for example **international**. However, some people can move for a short period of time such as **holiday in America**. Other people tend to move for a much longer period of time, e.g. **new job in south-east England**.
2 Forced, Voluntary, Forced, Voluntary.
3 Aim to include an example.
  Forced migration is the movement of people who have no choice. One example of this is people moving as a result of religious persecution as is the case of the Tutsi and Hutu people in Congo and Rwanda. People move because their life has been threatened. Once the threat is over, often people will return.
  Other examples – fleeing from war or invasion (Tibetans escaping Chinese troops) or natural hazards – Chances Peak eruption in Montserrat in June 1997.
4 Net in-migration is when there are more immigrants than emigrants, whereas with net out-migration there are more people leaving the country (emigrants) than coming in (immigrants).

104. Migration into and within Europe

1 Illegal immigrants from Africa, retirement migration from northern Europe.
2 (a)

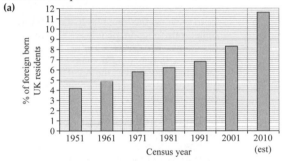

  (b) You need to refer to data from the graph.
      Overall trend increases, relatively steady growth from 1951–91, from just over 4 per cent in 1951 to over 11 per cent in 2010. Largest growth rate is from 2001–2010 est. with a predicted growth of 5 per cent.
      Other trends can be credited.
3 (a) A non-citizen of a country that has entered without proper sanction or continues to stay in the country after any visa has expired.
  (b) Examples include, e.g. Algeria or Libya.
      There are lots of illegal immigrants coming to the EU from northern parts of Africa such as Algeria and Libya. This can be voluntary migration due to economic reasons, such as lack of jobs and income in their own country, or more commonly from forced migration due to fighting, such as in Sierra Leone or the Democratic Republic of Congo.

105. Impacts of migration

1 Your answer will depend on your examples. Make sure your answer is well-explained and give a range of specific, explained points with relevant examples to support your points. Four marks are also available for SPaG, so make sure that your answer is clear and well-organised and that your spelling, punctuation and grammar are really good. A maximum of 2 marks can be given without some development of at least 1 point. Can be viewed from either positive and / or negative impacts. Must include both economic and social for full marks.
  Social: discrimination, racial tension, formation of ghettos, immigrants add to culture of host country.
  Economic: bring new skills which add to the economy, meets any shortage of workers, will add to unemployment levels during recessions, strain on housing, schools, health care, etc.
2 The host country is the country which the migrants move to whereas the country of origin is the country they have left behind.

3 The focus is on negative impacts so no credit can be given for positive comments. Aim to refer to an example.
The younger generation tend to move away and this causes a decrease in the birth rate. It is often the enterprising that move and this results in a loss of skilled workers and results in the brain drain. Fewer economically active left to support remainder of the population, often left with an ageing population which puts stress and pressure on services.

106. Factors behind migration

1 A push factor is one which forces you to move whereas a pull factor will encourage you to move.
2 Pull, Push, Push, Pull.
3 Three valid examples. Internet, email, texting, mobile phone calls, video conferencing, television.
4 The rate of illegal immigrants will decrease as it is made more difficult to come into the country.
5 Your answer will depend on your examples. Make sure your answer is well-explained and give a range of specific, explained points with relevant examples to support your points. These will be from different examples. Four marks are also available for SPaG, so make sure that your answer is clear and well-organised and that your spelling, punctuation, grammar and use of terminology are really good. A maximum of 2 marks can be given for descriptive comments. Example and explanation of one form of transport is needed for full marks.
With the introduction of high-speed trains such as the bullet train in Japan, people are able to travel much quicker than usual. The Channel Tunnel has increased accessibility to the Continent, in particular France. The introduction of fast jumbo jets such as the A380 has meant that places far away such as Australia and New Zealand are much more easily accessible.

107. Temporary migration

1 Short-term population flow means there is no permanent change of address.
2 Temporary migration.
3 Due to the poorer economic conditions in Eastern Europe, such as Poland and Czech Republic, many migrants moved for economic reasons – jobs and money. Many Eastern European countries joined the EU in 2004, this meant that their citizens could now move freely between countries.
4 Often combined treatment with a holiday, lower cost, waiting lists were smaller.
5 Example needed.
Footballers attracted by global reputation of English Premiership, attracted by high wages, glamorous lifestyle. Many events take place in different countries and so have to move to compete, e.g. golf, motor racing or tennis.

108. Retirement migration

1 Many elderly people are moving. This is called **retirement** migration. Elderly people often want a **quiet** environment and their houses tend to be **quite large** for their needs, so they downsize. Sometimes, elderly people **sell** their homes to help support their pension.
2 Population pyramid for the area can become top heavy, increased need for geriatric health care puts pressure on local health services and funding, more need for specialist accommodation, increased language barrier if the move is to a foreign country.
3 Attracted by warmer climates, e.g. Spain, want to move to more pleasant surroundings. Calmer or quieter area to be near friends / family. No longer the need to live near place of work, downsizing to smaller property, sell house to supplement pension.
4 Positive impacts – increased employment opportunities in geriatric sector.
Negative impacts – top-heavy population structure, pressure increased on health, welfare and social services, specialist accommodation needed, high percentage of elderly on the poverty line.

109. Types of 'grey' migration

1 Local, regional and international.
2 Smaller apartments, therefore more manageable to look after, often fitted with facilities for the elderly, e.g. lifts, mobility devices, central alarms, often located in popular, attractive areas such as Eastbourne.
3 (a) A grey resort is a specific area that attracts large numbers of ageing dependents.
   (b) Yarmouth, Worthing and Lytham St Anne's.
4 Aim to use map evidence.
Uneven distribution, east and south coast tend to be more popular. Central regions do not attract large numbers except La Rioja. Retirement migration is largest in south Valencia with +7.5 per cent. Other exception to pattern is South Galicia with 2–5 per cent increase.

110. Types of tourism

1 False, True, True, True.
2 Leisure – festivals, city break.
VFRs – parents for Christmas, staying with friend for weekend.
Business – conferences, meetings.
3 Accurately completed bars for the VFRs and Business data. One additional mark for completion of the key.

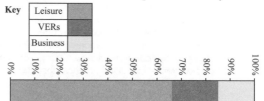

4 The state of being seasonal or dependent on the seasons of the year.

111. Why tourism is growing 1

1 (a) Accurate completion of the graph – only allow 1 mark if the line has been drawn without a ruler.

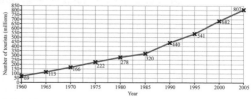

   (b) 209 million.
   (c) 1990–2000.
2 Any one from: increased leisure time, access to communication and IT, changing fashions.
3 Shorter working week (50 hrs/week in 1950s compared to 35–40 hrs today) so people have more free time to go on holiday. People are retiring earlier therefore more available time to go on holiday.

112. Why tourism is growing 2

1 Border controls relaxed = P, changes in exchange rates = E, journey times reduced = E, government change = P.
2 People nowadays have more disposable income (plus data), the minimum wage has increased (plus data), this has meant that people have more money to spend on holidays, fewer children in the family (plus data), this therefore increases the disposable income available and so people can go on more holidays.
3 Border controls have been relaxed, free movement of tourists, e.g. EU reform visas and departure tax provided economic support for the country, political reform, e.g. dissolution of the Soviet Union, implementation of government policies to help increase tourism investment from economic providers, e.g. world bank or IMF (plus data).

113. Tourist destinations

1 A form of activity holiday which involves things such as bungee jumping or snorkelling.
2 Any combination of attractions – 2×2, 1×3 or 3×1.
Human attractions – accommodation, transport links.
Physical attractions – steep slopes, beautiful scenery, forest, good weather.

3 Emphasis on change from hiking holiday in the summer to skiing – the slopes would be covered with snow, this would attract skiers and snowboarders, ski lifts would be in operation.

4 Holiday-makers are grouped together in a hotel or a complex of hotels containing all the services thought to be necessary for their sustenance and entertainment.

114. The Butler model: Blackpool

1 Top of table down – decline, exploration, consolidation, rejuvenation, development.

2 For Blackpool these could include: Blackpool Tower, North and South Pier, promenade, arcades and amusements.

3 Improved accessibility to many resorts, especially coastal such as Blackpool or Brighton. Car ownership was generally very low, encouraged many people to travel on trains.

4 Your answer will depend on your examples. Make sure your answer is well-explained and give a range of specific, explained points with relevant examples to support your points, these should be clearly linked to the case study of Blackpool. Four marks are also available for SPaG, so make sure that your answer is clear and well-organised and that your spelling, punctuation, grammar and use of terminology are really good. Max 3 marks if no reference to case study detail.
Blackpool was still highly popular as a seaside resort up until the end of WWII, but the introduction of the package holiday meant that people could visit places such as the Mediterranean more easily. As a result of this Blackpool started to go into decline. It attempted to stabilise by building the Sea Life Centre but still declined. Rejuvenated by rebranding in 2003 such as the new art centre in South Shore.

115. The Butler model: Benidorm

1 Exploration – 3 S's.
Involvement – Mayor allowed bikinis.
Development – Many hotels and restaurants opened.
Consolidation – Tourist numbers boomed.
Stagnation – Other holiday destinations became popular, e.g. Turkey.

2 In 1954, Benidorm was a town whose main function was fishing. The attractions for visitors were very much limited to the 3 S's – sea, sand and sun. This was stage 1.
The mayor campaigned to allow bikinis so it could encourage people seeking beach holidays to come and keep up with the fashion from the States. This was stage 2.
The 1960s show hotels and restaurants being developed and in the 1970s infrastructure was improved. This was stage 3.

3 Consolidation or stage 4.

4 Aim to include an example and ensure that decline and rejuvenation are covered.
In the 1990s in Benidorm, there was a considerable drop in hotel occupancy with only 71 per cent of places taken, this was the start of the decline phase. In 1995, Benidorm attempted to rejuvenate by building theme parks such as Terra Mítica which opened in 2001 – this encouraged tourists by offering a different experience.

116. Social impacts of tourism

1 Can be either negative or positive – need two examples. Local crafts can be sold, opportunity to mix cultures, ignoring dress code, loud and aggressive behaviour from tourists, encourages prostitution.

2 Provision of jobs and income by selling of local crafts (plus example). Increased entertainment, e.g. Kukyekukyeku tribe in Ghana, reduces stereotypes, provides improvement in infrastructure, e.g. health and education (plus example).

3 Culture clash, e.g. strict dress codes in many foreign countries such as Turkey; loss or dilution of local culture; seasonal unemployment (plus detail), increase in second homes which forces locals out as too expensive to buy (plus example).

117. Economic and environmental impacts of tourism

1 Economic – more jobs are created, increased wealth of countries, leakage abroad, seasonal unemployment. Environmental – pollution of water sources, protection of endangered animals, loss of animal habitats.

2 Increase in local employment, e.g. Lake District in Cumbria 31 000 people are employed within the tourist industry. Increase in average earnings, e.g. excess of £900 million per year in Lake District National Park. People can be educated about wildlife and preserving natural habitats in eco-tourism regions (plus example), profits made can be put back into protecting value areas or improving the infrastructure of the surrounding region.

3 Habitats are cleared; overuse of water supplies; rubbish and sewage produced; coral reefs destroyed, e.g. Great Barrier, Australia; wildlife disturbed; traffic congestion; air pollution; noise pollution. Aim to give explanation.

118. Eco-tourism

1 Eco-tourism is a form of tourism that tries to **minimise** the damage done to the **environment**. It also aims to provide **jobs** for local people and ensures that the **profits** made stay within the area. It is a form of **eco-friendly** tourism.

2 Your answer will depend on your examples. Make sure your answer is well-explained and give a range of specific, explained points with relevant examples to support your points. Four marks are also available for SPaG, so make sure that your answer is clear and well-organised and that your spelling, punctuation and grammar are really good. Needs to refer to local people. Example required for full marks.
In the Amira Valley in Trinidad, local people have access to jobs and training, interest-free loans are on offer, food is supplied by local growers.

3 Must refer to **environmental** benefits, not people unless related or linked back. Example must also be given.
For example in the Asa Wright Nature Centre, they have only allowed 10 per cent of the land to be available to tourists. This discourages mass tourism and therefore limits the impacts on the environment. On-site recycling of refuse and use of wastewater helps to preserve the environment, minimal number of trails (only 5 designated), any profits made go back into conservation and preservation of the nature centre.

Published by Pearson Education Limited, Edinburgh Gate, Harlow, Essex, CM20 2JE.

www.pearsonschoolsandfecolleges.co.uk

Copies of official specifications for all Edexcel qualifications may be found on the Edexcel website: www.edexcel.com

Text and original illustrations © Pearson Education Limited 2013
Edited, produced and typeset by Wearset Ltd, Boldon, Tyne and Wear
Illustrated by Wearset Ltd, Boldon, Tyne and Wear
Cover illustration by Miriam Sturdee

The rights of Anne-Maire Grant to be identified as author of this work have been asserted by her in accordance with the Copyright, Designs and Patents Act 1988.

First published 2013

16 15 14 13
10 9 8 7 6 5 4 3 2 1

British Library Cataloguing in Publication Data
A catalogue record for this book is available from the British Library

ISBN 978 1 446 90538 7

Printed in Slovakia by Neografia

Acknowledgements
The publisher would like to thank the following for their kind permission to reproduce their photographs:

(Key: b-bottom; c-centre; l-left; r-right; t-top)

Alamy Images: geogphotos 24, imagebroker 32, incamerastock 89b, Justin Kase 39, Kevin Schafer 44, Mike Greenslade 1, Robert Harding World Imagery 33, Tomislav Zivkovic 45, Tremorvapix 21; DK Images: Nigel Hicks 38, Tony Souter 89t; Environment Agency copyright. All rights reserved: 37; Getty Images: Visit Britain/Martin Brent 23; Shutterstock.com: Vilant 113; Veer/Corbis: billperry 49

All other images © Pearson Education Limited

We are grateful to the following for permission to reproduce copyright material:
Figures
Figure on page 4 from Maps International, Ordnance Survey on behalf of HMSO, © Crown Copyright 2013. All rights reserved. Ordnance Survey Licence number 100030901; Figure on page 2 from U.S. Population Density (By Counties), http://www.census.gov/dmd/www/pdf/512popdn.pdf, United States Census Bureau; Figure on page 15 from Transport for London, Crown Copyright Open Government Licence; Figure on page 20 from http://revisionworld.co.uk/a2-level-level-revision/geography/coastal-environments/other-factors/coastal-geology; Figure on page 56 from EEA 2001 (European Economic Association); Figure on page 64 from http://www.iwmi.cgiar.org/Publications/IWMI_Research_Reports/PDF/PUB123/RR123.pdf, Reproduced by kind permission of International Water Management Institute (IWMI); Figure on page 76 from http://www.climate-connect.co.uk/Home/?q=node/1864 Climate Connect, Source: Climate Connect Ltd; Figure on page 84 from http://www.scotland.gov.uk/Publications/2004/01/18714/31269, The Scottish Government Crown Copyright, Open Government Licence; Figure on page 100 from http://www.geographylwc.org.uk/A/AS/ASpopulation/popstructure1.html, Lord Wandsworth College, Graham Smith, Head of Geography

Maps
Maps on page 2, 3, 4, 8 from Maps International, Supplied by courtesy of Maps International and Ordnance Survey on behalf of HMSO, © Crown Copyright 2013. All rights reserved. Ordnance Survey Licence number 100030901; Map on page 36 after http://www.guardian.co.uk/environment/2010/jan/29/cost-of-uk-flood-protection, Guardian News and Media Ltd; Map on page 39 from http://www.bgs.ac.uk/research/earthquakes/newZealandFeb2011.html, British Geological Society; Map on page 45 from http://www.ga.gov.au/ausgeonews/ausgeonews200806/disaster.jsp, Alanna Simpson, Phil Cummins, Trevor Dhu, Jonathan Griffin and John Schneider, Creative Common Licence; Map on page 52 from http://epp.eurostat.ec.europa.eu/portal/page/portal/waste/key_waste_streams/waste_electrical_electronic_equipment_weee Eurostat; Adminstrative boundaries © EuroGeographics © UN-FAO © Turkstat; Cartography: Eurostat – GISCO 10/2012, Source: Eurostat, http://epp.eurostat.ec.europa.eu, © European Union, 1995–2013; Map 2.15 from http://www.theclimatehub.com/uk-wind-farm-capacities-map, Department of Energy and Climate, Crown Copyright/Open Government Licence; Map on page 69 from http://news.bbc.co.uk/1/hi/uk/7971257.stm, Ofwat/Crown Copyright/Open Government Licence; Map on page 96 from http://www.ons.gov.uk/ons/search/index.html?pageSize=50&sortBy=none&sortDirection=none&newquery=population+density+london+2009, Reproduced from 2009 Ordnance Survey map with the kind permission of the Ordnance Survey; Map on page 109 from http://www.european-citizenship.net/details.php, INE 2005 IFH 2006 Dr. Michael Janoschka, Creative Commons Licence

Text
Extract on page 33 from *AQA Geog short course Spec A June 2011 FT (40304)*, AQA examination question reproduced by permission of AQA

Every effort has been made to contact copyright holders of material reproduced in this book. Any omissions will be rectified in subsequent printings if notice is given to the publishers.

In the writing of this book, no Edexcel examiners authored sections relevant to examination papers for which they have responsibility.